STUDENT HANDBOOK
QUESTIONS, PROBLEMS, AND SOLUTIONS

GENES VIII
LEWIN

ROBERT J. GREBENOK • THOMAS LEUSTEK • HARRY NICKLA
Canisius College   *Rutgers University*   *Creighton University*

GREGORY J. PODGORSKI • MARK REEDY • DENNIS L. WELKER
Utah State University   *Creighton University*   *Utah State University*

PEARSON

Prentice Hall

Upper Saddle River, NJ 07458

Editor-in-Chief, Science: John Challice
Executive Editor: Gary Carlson
Project Manager: Crissy Dudonis
Vice President of Production & Manufacturing: David W. Riccardi
Executive Managing Editor: Kathleen Schiaparelli
Assistant Managing Editor: Becca Richter
Production Editor: Elizabeth Klug
Supplement Cover Manager: Paul Gourhan
Supplement Cover Designer: Joanne Alexandris
Manufacturing Buyer: Ilene Kahn
Image Credit : High Density Liquid Crystalline DNA /National High Magnetic Field
Laboratory, Florida State University

© 2004 Pearson Education, Inc.
Pearson Prentice Hall
Pearson Education, Inc.
Upper Saddle River, NJ 07458

Printed in the United States of America

10   9   8   7   6   5   4   3   2   1

ISBN   0-13-146411-6

Pearson Education Ltd., *London*
Pearson Education Australia Pty. Ltd., *Sydney*
Pearson Education Singapore, Pte. Ltd.
Pearson Education North Asia Ltd., *Hong Kong*
Pearson Education Canada, Inc., *Toronto*
Pearson Educación de Mexico, S.A. de C.V.
Pearson Education—Japan, *Tokyo*
Pearson Education Malaysia, Pte. Ltd.

Contents

Introduction: Students: Read This Section First!!!   Page   1

Chapter			
	1	Genes are DNA	7
	2	The interrupted gene	17
	3	The content of the genome	26
	4	Clusters and repeats	34
	5	Messenger RNA	43
	6	Protein synthesis	54
	7	Using the genetic code	63
	8	Protein localization	73
	9	Transcription	86
	10	The operon	97
	11	Regulatory circuits	109
	12	Phage strategies	117
	13	The replicon	124
	14	DNA replication	129
	15	Recombination and repair	140
	16	Transposons	151
	17	Retroviruses and retroposons	161
	18	Rearrangement of DNA	171
	19	Chromosomes	182
	20	Nucleosomes	188
	21	Promoters and enhancers	198
	22	Activating transcription	208
	23	Controlling chromatin structure	218
	24	RNA splicing and processing	227
	25	Catalytic RNA	236
	26	Immune diversity	244
	27	Protein trafficking	252
	28	Signal transduction	260
	29	Cell cycle and growth regulation	268
	30	Oncogenes and cancer	275
	31	Gradients, cascades, and signaling pathways	282

Students: Read This Section First!!!

The intent of this collection of questions, problems and solutions is to help you understand the specifics, comparative aspects and concepts of molecular biology as presented in the text *Genes VIII* by Benjamin Lewin.   Take comfort in knowing that what at first appears to be an overwhelming assortment of unrelated facts, through study, common themes will emerge around which specifics will organize. Because all organisms accomplish the task of living in a fairly universal manner, common themes and subtle variations will be the rule. By focusing on common themes, and appreciating variations on those themes, one can manage a fairly sophisticated understanding of molecular biology. There is no magic formula for understanding complex material, but there are strategies which increase the likelihood of success. Some of those strategies are presented below.

Structure of Genes VIII and Strategies for Success

General Structure:

Table of Contents: Notice the hierarchy of subjects, beginning with the smallest unit *Genes* and ending with *Cells*. Because of the interconnectedness of the material, if your instructor does not assign every chapter for your class, read the introduction and summary for those chapters not assigned. This will provide you with a more global view of the common themes and more easily allow you to provide an appropriate context for your assigned chapters.

Main Body: 31 chapters, each containing an Introduction, Sections, Summary, and References.

Glossary: pages 981-1002. Look up words you don't know!

Index: pages 1003-1027.

Access to www.prenhall.com/lewin   A rather unique feature of the *Genes VIII* text is its relationship to a fairly sophisticated, searchable, on-line version of the text. Just inside the cover is a page which provides simple directions for registration and login. The registration process is simple and takes approximately 10 minutes. Once you have obtained a login and password name, you have access to a variety of resources which include the following:

Useful **search engine** which directs you to appropriate sections of *Genes VIII* based on your query.

Key Concepts and **Key Terms** are presented for each section of each chapter.

Links (left side of screen) which directs you to other texts which relate to your query.

Chapter details are provided with text figures to which your can **add your own notes** and cross references. This is a very useful feature which allows you to annotate the text. In addition, your entries are maintained for future reference. In other words, when you leave and return at a later date, your notes are maintained in the database.

Students: Read This Section First!!!

Chapter Layout and Your Approach:

The Introduction sets the general stage of what is to come and links each chapter to preceding information. Read the Introduction first, then skip to the Summary at the end of the chapter. However, confusing, read the Summary and notice that considerable detailed information is presented. After reading the Summary, re-read the Introduction, then begin reading each individual assigned section. If your instructor assigns only certain chapters in the text, it may be helpful to read the Introductions of those chapters not assigned.

Key Concepts: Common themes for each section are delivered in boxes entitled Key Concepts. For example, Section 3.15 on page 72 is entitled *Genes are expressed at widely differing levels* and below it is a box containing the following concepts.

- In any given cell, most genes are expressed at a low level.
- Only a small number of genes, whose products are specialized for the cell type are highly expressed.

Following these Key Concepts is a section approximately one page long with detailed information which includes the following related information:

- Saturation analysis identifies the percentage of template DNA for mRNA
- The number of expressed genes in yeast and higher eukaryotes
- Exception is human brain
- Kinetic analysis and sequence complexity (Rot values) (Figure 3.31)
- Proportions of mRNA species
- Abundance and the average number of mRNA molecules for each species
    - Abundant mRNA = 1000-10,000 copies per cell
    - Scarce or complex mRNA = <10 copies per cell

The above structure is at the heart of **Genes VIII** and mastery will depend on your ability to confront highly complex and, depending on the organism, seemingly conflicting information. Each page will typically contain three to four concepts, several entirely new terms, and multiple examples. So, how do you avoid the following?

"I studied all the material but failed your test."

"This is the first test I have **ever** failed."

"I helped three of my friends last night and I got the lowest grade."

"I stayed up all night studying for your exam and I still failed."

Students: Read This Section First!!!

Study when there are no tests!

1. Set disciplined study goals.
2. Study when there are no tests.
3. Read assigned chapters before class.
4. Attend class.
5. Review <u>Key Concepts.</u>
6. Do section-related problems in this book.
7. Strive for common themes, note comparisons, and expect variations.
8. Be careful with old exams.
9. Don't "second guess" the instructor.

Don't cram. A successful tennis player doesn't learn to play tennis overnight; therefore, you can't expect to learn molecular biology under the pressure of night-long cramming. It will be necessary for you to develop and follow a realistic study schedule for this subject as well as the other courses you are taking. It is important that you focus your study periods into intensive, but short sessions each day throughout the entire semester (quarter, trimester). Because tests often require you to think "on the spot" it is very important that you get a good night's sleep before each test. Avoid caffeine in the evening before the test because a clear, rested, well-prepared mind will be required.

> ***Study when there are no tests***

Study goals. The instruction of molecular biology is often divided into large conceptual units. A test usually follows each unit. It will be necessary for you to study on a routine basis long before each test. To do so, set specific study goals. Adhere to these goals and don't let examinations in one course interfere with the study goals of another course. Notice that each course being taken is handled in the same way --- study ahead of time and don't cram.

Study each subject at least every other day --- especially when there are no tests!

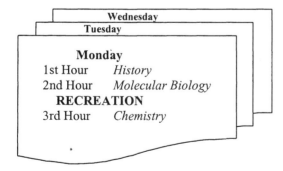

> Schedule each day in each month; schedule each month in the semester

Attendance and Attention
Are Mandatory

Many professors do not take attendance in lectures; therefore, it is likely that some students will opt to take a day off now and then. Unless those students are excellent readers and excellent students in general, continual absences will usually result in failure. While using the text is important in your understanding of molecular

> *Since it is the instructor who writes and grades the tests, who is in a better position to prepare the students for those tests?*

biology, the instructor can walk you through the concepts and strategies much more efficiently than a text because a text is organized in a sequential manner. A good teacher can "cut and paste" an idea from here and there as needed. To benefit from the wisdom of the instructor, the student must concentrate during the lecture session rather than sit, passively taking notes, assuming that the ideas can be figured out at a later date. Too often the student will not be able to relate to notes passively taken weeks before. In addition, the instructor will not be able to cover all the material in the text. Parts will be emphasized while other areas may be omitted entirely.

How to Think About Molecular Biology

There is no magic formula for understanding molecular biology or any other discipline of significance. Learning anything, especially at the college level, requires time, patience, and confidence. First, a student must be willing to focus on the subject matter for an hour or so each day over the entire semester (quarter, trimester, etc.). Study time must be free of distractions and pressured by realistic goals. The student must be patient and disciplined. It will be necessary to study when there are no assignments due and no tests looming.

The majority of successful students are willing to read the text ahead of the lecture material, spend time thinking about the concepts and examples, and work as many sample questions and problems as possible. They study for a period of time, stop, then return to review the most difficult areas. They do not try to cram information into marathon study sessions a few nights before the examinations. While they may get away with that practice on occasion, more often than not, understanding the specifics and concepts of molecular biology requires more mature study habits and preparation.

Read ahead. You have been told that it is important to read the assigned material before attending lectures. This allows you to make full use of the information provided in the lecture and to concentrate on those areas that are unclear in the readings. Depending on each classroom circumstance, an opportunity is often provided for asking

> ***Ask Questions***
> ***and***
> ***Don't Tune Out!***

questions. Your questions will be received much more favorably if you can say that after reading the book and listening to the lecture a particular point is still unclear. It is very likely that your question will be quickly dealt with to your benefit and the benefit of others in the class.

Students: Read This Section First!!!

Molecular biology as presented in *Genes VIII* is strong on examples, comparisons and processes which interact in a variety of ways. Models describe the manner in which genes are constructed and how they function. Because many parts of the models interact both in time and space, molecular biology cannot be viewed as a discipline filled with facts which should be memorized. Rather, one must be, or become, comfortable with seeking to understand not only the specifics of the models but also how the models exist, with variation, among organisms. For example, a gene in Eubacteria, Archaea, and Eukaryotes is transcribed in a similar manner, however, the genes themselves have structures which can be viewed as variations on a common themes. While transcription produces RNA as a 5'-3' product in all organisms, most genes in Eukaryotes occur in pieces which is not so in the other groups mentioned above.

There are no practice questions and problems in Genes VIII, thus the purpose of this book!

To help you adjust to the variety of examples and approaches to concepts, instructors often assign practice problems from which you gain some appreciation of what the instructor expects you to comprehend from each chapter. Since there are no questions or problems in *Genes VIII*, we developed this book to help you assess your mastery. If your instructor has assigned certain problems from this book, finish working all of them *at least* one week before each examination. Before starting a set of problems, read the respective chapter carefully and consider the information presented in class. Follow the suggestions in the box to the right.

Expect to make mistakes and learn from those mistakes. Sometimes what is difficult to see one day may be obvious the next day. If you are still having problems with a concept, schedule a meeting with your instructor. Usually the problem can be cleared up in a few minutes.

You will notice that at the end of each chapter in this book, we have presented the solution to each question. For test purposes, instructors may take questions directly from those presented here or they may modify them. Reversing the "direction" of a question is a common approach. As you work through the solutions, consider different ways in which the same concept can be addressed by a slightly different wording of a question.

Suggestions for working the questions and problems:

(1) Work each without looking at the answer.

(2) Check your answer at the end of each chapter in this book.

(3) If incorrect, work the problem again.

(4) If still incorrect, you don't understand the concept.

(5) Re-read your lecture notes and the text.

(6) Work the problem again.

(7) If you still don't understand the solution, mark it, and go to the next problem.

Be careful when using old examinations. Often it is customary for students to request or otherwise obtain old examinations from previous students. Such a practice is loaded with pitfalls. First, students often, albeit unconsciously, find themselves "second guessing" about questions on an upcoming examination. They forget that an examination usually only tests over a subset of the available information in a section. Therefore entire "conceptual areas" may be available which have not appeared on recent exams.

Often the reproductions of old examinations are of poor quality (having been copied and passed around repeatedly) and it is difficult to determine whether the answer provided is correct. In addition, if a question has the same general structure as one on a previous examination, but is modified, students often provide an answer for the "old" question rather than the one being asked. Granted, it is of value to see the format of each question and the general emphasis of previous examinations, but remember that each examination is potentially a new production capable of covering areas which have not been tested before. This is especially likely in a course such as molecular biology where the material changes very rapidly.

> *Don't try to figure out what will be asked.*
> *Study all the material as well as possible.*

Structure of This Book

The intent of this book is to help you understand molecular biology as given in the ***Genes VIII*** and most likely in the lectures. Rather than merely provide you with the solutions to the problems, we have tried to give you a fairly detailed solution (especially to the more complex questions) so that you can see where information is obtained and how it can be applied in the solution. At the beginning of each chapter is a section which lists general concept areas and related questions in the following groupings:

- **Section A**: Recall and short answer

- **Section B**: Analytical and critical thinking

- **Section C**: Extensions and applications

- **Section D**: Standardized test format

Depending on the particular goals and level of your course in molecular biology, your instructor may have you concentrate on one section at the expense of another. However, to fully test your understanding, you should attempt all the questions in each section. Short of that, we suggest that you attempt to answer as many questions as possible from each General Concept and/or Processes/structures section as indicated in the shaded box at the beginning of each chapter.

Chapter 1

Genes are DNA

General Concepts	Section	Related Questions
Genes and DNA	Intro. – 1.4	A1, A2, A3, A4
DNA structure	1.5-1.6	A6, A7, A9, B1, B3, C2, D1, D3
DNA replication	1.8-1.9	A8, B2
Mutation	1.10-1.18	A10, A11, A12, D5, D7, D8, D9, D10
Wild type	1.19	A15
Recombination	1.20	
Genetic code	1.21-1.23	C2
Colinearity	1.23	D5, D6, D10
Gene expression	1.24-1.26	A5, A13, B5, C1, D4
Hereditary agents	1.27	A5
Processes/structures		
Transformation	1.2	A4
Mutation/reversion	1.12	A10, A11, A12, D5, D6, D7, D8, D9, D10
Modified bases	1.14	A12
Complementation	1.16	A14
Transcription	1.24	B5, D4
Translation	1.25	A2
Trans- and *cis*-acting	1.25	C1
Examples, Applications and Techniques		
Differential labeling	1.3, 1.7	A4, A8, B2, B3, D2
Nucleic acid hybridization/denaturation	1.9	A9, B4, B5

Questions:

A. Recall and short answer

1. Physically, the genome is divided into a number of different nucleic acid sequences, whereas functionally, the genome is divided into a number of _____.

2. A gene is a sequence of DNA that, by transcription, codes for a _____; in protein-coding genes, the _____ in turn codes for a _____.

3. Describe the relationship between the process of transformation in *Pneumococci* and the discovery that DNA is the genetic material in bacteria.

4. Present an overview of two classical experiments which demonstrated that DNA is the genetic material in bacteria and viruses. Can RNA be the genetic material?

5. Describe four major functions of DNA in a prokaryotic or eukaryotic cell.

6. The composition of DNA can be described in a number of ways, one of which is the "GC" content which varies from about 26% to 74% among different species. If the GC content of a DNA molecule is 40%, what are the molar percentages of the four bases (G, C, T, A)? If the DNA is single-stranded, would you change your answer?

7. Double-stranded nucleic acids are said to be antiparallel. What structural configuration is antiparallel?

8. The Meselson and Stahl (1958) experiment provided direct evidence for the semiconservative replication of DNA in *E. coli*. What density pattern of bands would occur in a centrifugation gradient for semiconservative replication after one generation of replication of parental DNA (^{15}N) in ^{14}N medium?

9. The melting (denaturation) temperature of DNA covers a fairly narrow temperature range, the midpoint of which is symbolized as T_m. It depends on the proportion of GC base pairs such that nucleic acids with a higher GC content have a higher T_m. Why?

10. The most common class of point mutation is called a *transition mutation* which comprises the substitution of one _____ by another _____ or one _____ by another _____.

11. BrdU (bromouracil) is a mutagen which causes mispairing when incorporated into DNA because of its ambiguous pairing properties. It is incorporated in the place of thymine but in its enol form can base pair with guanine. Following replication, the original AT pair is replaced by a GC pair. Which class of mutation, *transition* or *transversion*, would be caused by BudU?

12. The consequences of deamination are different for 5-methylcytosine compared with cytosine. Deaminating the relatively rare 5-methylcytosine causes a mutation whereas deamination of the more common cytosine usually leaves the DNA free of mutation. Why the difference?

13. Provide a brief overview of the *one gene: one enzyme hypothesis*.

14. Failure of two mutations to complement (produce the wild type phenotype) when they are present in the *trans* configuration in a heterozygote indicates that the two mutations are in the _____ (*same* or *different*, state which) gene.

15. Is it possible that there can be more than one wild-type allele for a given locus? Explain and give an example.

B. Analytical and critical thinking

1. Consider the structure of double-stranded DNA. When DNA is placed into distilled water it denatures, that is, the two strands separate; however, by adding NaCl to the mixture, the DNA renatures. Why?

2. Assume that a strain of *E. coli* is grown for many generations in ^{15}N, a heavy isotope of nitrogen (^{14}N). Assume also that the DNA has a density of 1.720 gm/cm^3 (water is 1.00 gm/cm^3) and that DNA containing only ^{14}N has a density of 1.700. In an experiment similar to that of Meselson and Stahl, bacteria from the ^{15}N culture were washed in buffer and transferred to ^{14}N medium for one generation and DNA was extracted.

(a) What would be the expected density of the extracted DNA or DNAs ?

(b) Assuming that the extracted DNA was completely denatured (heat at 90° C for 15 minutes), what would be the density (or densities) of the DNA (or DNAs) in the heated extract? Explain your answer. (Note: although not actually the case, assume that DNA has the same density whether single- or double-stranded.)

(c) Assuming that the molar percentage of adenine in the extracted DNA was 30%, what would be the expected molar percentages of the other nitrogenous bases in this DNA?

3. Assume that the following tetranucleotide was synthesized from typical nucleotides except that the innermost phosphate (the one attached to the 5' carbon) of the cytosine nucleotide was labeled with radioactive phosphate (^{32}P). The tetranucleotide was digested with the enzyme spleen phosphodiesterase which cleaves between the phosphate and the 5' carbon. Present a diagram of the resulting ^{32}P-labeled nucleotide.

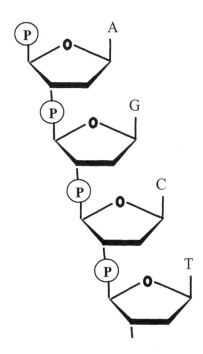

4. A common and simple procedure for calculating the melting temperature for relatively short DNA segments under *in vitro* (test tube) conditions is to apply the following formula:

$$T_m \, (°C) = 81.5 + 0.41(\%GC) - (675/N)$$

where %GC is the percentage of G and C nucleotides in the DNA and N is the length of the DNA given in nucleotides. Calculate the T_m for the following DNA segment noting that only one of the two complementary strands is provided.

5'-TTGTGAATGCTGGCCATCTC-3'

5. Antisense oligodeoxynucleotides are relatively short single-stranded DNAs that may selectively block disease-causing genes by base pairing with RNA transcripts. The RNA/DNA hybrid is degraded by a specific intracellular enzyme. Cancer genes have often been chosen as potential targets for antisense drugs, however, some data indicate that RNAs other than the ones intended are subject to antisense elimination (Cho et al. 2001 *PNAS* 98: 9819-9823). (a) Why might one see non-related gene expression decreased by the introduction of antisense oligodeoxynucleotides into a cell? (b) What drawbacks might one expect in the use of antisense therapy for genetic diseases?

C. Extensions and applications

1. Common features of DNA structure, replication and function among all organisms suggests that some DNA sequences may be common among organisms as well. In addition, since various sections of DNA function similarly in terms of protein synthesis one might expect that homologous regions in different genes of the same organism or in genes of different organisms might be similar. Such conserved sequences have been identified and are known as the Pribnow box and -35 region in prokaryotes and the Golberg-Hogness region in eukaryotes. Work by Novitsky et al. (2002 *J. Virology* 76:5435-5451) indicates that among isolates of the Human Immunodeficiency Virus (causes AIDS), among other conserved sequences, a GGGNNNNNCC consensus sequence exists as a *cis*-acting regulatory element which is critical in HIV replication and infection. What is meant by a *cis*-acting sequence? In terms of developing HIV therapies, what advantages would you see in finding such consensus sequences?

2. A number of human neurodegenerative disorders including Huntington's disease, fragile X syndrome, and myotonic dystrophy are caused by stretches of repeated sequences of three bases each. Such "trinucleotide repeats" are members of a larger category of repeats of 1 to 6 bases (N_1 to N_6) called microsatellites. When Toth et al. (2000 *Genome Research* 10:967-981) examined the frequency of classes of microsatellites among genes of a variety of taxonomic groups, the distribution was far from random as indicated by the data below:

Organism	(N_1)	(N_2)	(N_3)	(N_4)	(N_5)	(N_6)
Fungi	9	4	318	2	35	219
Yeast	36	19	706	7	52	330
Primates	49	10	1126	29	57	244

(a) What general conclusions are apparent from the above data?

(b) What factor(s) might account for the distributions presented?

(c) Would you expect the same pattern of microsatellite frequencies in non-coding regions of DNA?

D. Standardized test format

1. The basic structure of a nucleotide includes the following components

    A. amino acids, proteins.
    B. carbohydrates and lipids.
    C. base, sugar, phosphate.
    D. RNA and DNA.
    E. nucleic acids and lipids.

2. To identify the genetic material in bacteriophage, investigators in 1952 used differential labeling to determine that DNA is the genetic material. To do so, they labeled the phage with

    A. ^{32}P and ^{35}S.
    B. RNA and DNA.
    C. phosphates and sulfates.
    D. ^{15}N and ^{14}N.
    E. ethidium bromide.

3. Considering the structure of double-stranded DNA, what kinds of bonds hold one complementary strand to the other?

    A. covalent
    B. ionic
    C. Van der Waals
    D. hydrogen
    E. hydrophobic and hydrophilic

4. Which of the following accurately describes DNA as it exists in cells?

    A. Double-stranded, parallel, (A+G)/(C+T)=1.0
    B. Double-stranded, antiparallel, (A+G)/(C+T)=1.0
    C. Single-stranded, parallel, (A+G)/(C+T)=1.0
    D. Double-stranded, parallel, (A+G)/(C+U)=1.0
    E. Double-stranded, antiparallel, (A+G)/(C+T)=variable

5. The relationship between a protein-coding gene and a messenger RNA is that

    A. genes are made from mRNAs.
    B. mRNAs are made from genes.
    C. mRNAs make proteins which in turn make genes.
    D. all genes are made from mRNAs.
    E. none of the above.

6. Considering frameshift mutations generated by acridine dyes in the r_{II} region of a T phage, the addition or removal of three bases suggests that

    A. genes are read three amino acids at a time.
    B. coding frames are composed of six bases.
    C. there are three amino acids per nitrogenous base.
    D. there is a 1:1 coding ratio of bases to amino acids.
    E. the code is made up of three bases.

7. Considering frameshift mutations in the r_{II} region of phage, circumstances were observed in which the addition of a base followed by the removal of a base gave a wild type phenotype, whereas the removal of a base followed by the addition of a base gave a mutant phenotype. How can this apparent discrepancy be explained?

    A. There is no logical explanation for this observation.
    B. Each circumstance sets up a different combination of triplets in the "out-of-phase" region.
    C. Multiple "out-of-phase" regions are set up in the first circumstance and not the second circumstance.
    D. There is no "out-of-phase" region expected in the first circumstance.
    E. There is no "out-of-phase" region expected in the second circumstance.

8. Mutation suppression usually refers to a condition in which

    A. a mutation in one gene alters the expression of a mutation in a non-allelic gene.
    B. more than one mutation causes a mutant phenotype.
    C. one mutation enhances the mutant expression of a non-allelic mutant gene.
    D. every time a mutation occurs, physiological adjustments modify its influence on the phenotype.
    E. none of the above.

9. Two terms used to describe categories of mutational nucleotide *substitutions* in DNA are called

    A. base analogues and frameshift.
    B. frameshift and spontaneous.
    C. transversions and transitions.
    D. interactive and proactive.
    E. sense and antisense.

10. A class of mutations that results in multiple contiguous amino acid changes in proteins is probably caused by the following type of mutation.

      A. frameshift
      B. transversion
      C. transition
      D. base analogue
      E. recombinant

Answers:

A. Recall and short answer

1. genes

2. RNA, messengerRNA, protein

3. Transformation occurs when the DNA of one organism is genetically altered by exposure to DNA from another organism. Since DNA can carry heritable information from one organism to another, it must be the genetic material.

4. (1) The classic experiment demonstrating transformation in bacteria involved the exchange of genetic information from one *Pneumococcal* strain to another. In 1944, the transforming material was shown to be DNA. (2) Using differential labeling of proteins and nucleic acids in T2 phage, it was shown that the nucleic acid component (DNA) was the material which was injected into the bacterium during the lytic cycle. RNA is the genetic material in some viruses.

5. *Storage*: nucleic acids serve to store genetic information to allow stable maintenance and passage of that information. *Replication*: the genetic material makes copies when cells are duplicated. *Expression*: through the production of RNA and protein, the genetic material is expressed. *Variation*: the genetic material is capable of alteration by mutation.

6. G = 20%, C = 20%, A = 30%, T = 30%. The G+C content would still be be 40% and the A+T content would be 60% but there would be no base-pair restrictions.

7. The C-5' to C-3' orientations of the complementary strands run in opposite directions.

8. After one generation in the ^{14}N, there would be one hybrid-density, intermediate band.

9. Since GC pairs are held together with three hydrogen bonds rather than two, it takes greater energy to separate GC rich nucleic acids.

10. pyrimidine, pyrimidine .... purine, purine

11. transition because the substitution is pyrimidine to pyrimidine

12. The deamination of 5-methylcytosine produces thymine which then pairs with adenine in the next round of replication. Deamination of cytosine produces uracil which is usually removed through the action of uracil-DNA-gylcosidase. This results in an unpaired guanine residue which is open to a normal cytosine incorporation during repair synthesis.

13. Metabolic reactions are catalyzed by enzymes which are encoded by genes. A mutation in a gene alters the action of an enzyme thereby altering a metabolic pathway.

14. same; in order to complement, two mutations must be in different genes (or cistrons).

15. It is possible for a given locus to have multiple wild-type alleles. For example, with the *A* and *B* alleles which specify human blood types, neither can be considered mutant and are therefore described as polymorphic.

B. Analytical and critical thinking

1. In the DNA double helix, the negatively charged phosphates repel each other on the two sides of the helix. In distilled water, these charges are not neutralized (by positively charged ions) and the hydrogen bonds which hold the double helix together are broken. In the NaCl solution, the negatively charged phosphates are masked by the positively charged Na^+ ions, thus, the repulsion between the strands is lowered and the DNA molecule remains intact.

2. (a)   approximately 1.710 gm/cm^3
    (b)   1.720 and 1.700 gm/cm^3
    (c)   Thymine = 30%, Guanine = 20%, Cytosine = 20%

3. The labeled phosphate would now be attached to the 3' carbon of the 5' neighbor.

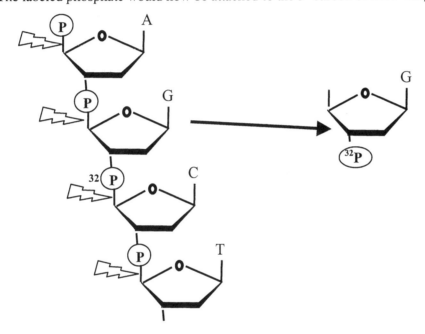

4. $T_m = 81.5 + 0.41(50) - (675/20)$

$T_m = 81.5 + 20.5 - 33.75$

$T_m = 68.25 \ °C$

5. (a) Since the antisense strand is relatively short, under physiological conditions, non-specific binding of the antisense strand to other RNAs in the cell may cause the degradation of non-targeted RNAs. (b) One of the major drawbacks to antisense therapy is the lack of specificity mentioned in part (a) and the unknown behavior of DNA oligodeoxynucleotides in living cells. In addition, such oligodeoxynucleotides would probably only be effective in "gain-of-function" mutations where a defective product causes the undesired phenotype.

C. Extensions and applications

1. A *cis*-acting sequence is usually represented by a contiguous stretch of DNA which is a target for RNA and/or protein interaction. *Cis*-acting elements are not themselves expressed as RNA or protein products. If sequence homologies can be identified for common strains of HIV, then it might be possible to develop vaccines which specifically target such sequences. Since the sequence is part of a regulatory site for an essential HIV gene, perhaps blocking that site would inhibit HIV replication. However, should a vaccine be developed for that specific site, it is possible that other HIV variants might escape vaccination.

2. (a) The frequencies of trinucleotides (N_3) and hexanucleotides (N_6) are similar in all three taxonomic groups, being higher than the other repeat motifs. (b) Clearly, there are more trinucleotide repeats than any of the other types. It is possible that the mechanisms which generate repeats favor tri- and hexanucleotide formations. It is also possible that selection differentially selects for some (tri- and hexa-) over others. Because the genetic code is triplet, inclusion of tri- and hexanucleotide repeats within a gene would maintain the proper reading frames and cause less dramatic alterations in protein structure. The answer therefore seems to be based on the non-overlapping nature of a triplet code. (c) Since genes are dependent on a triplet code for proper function and frameshifting would occur for N_1, N_2, N_4, and N_5 motifs, one would expect selection to reduce their frequency in coding regions but not in non-coding DNA sequences.

D. Standardized test format

1. C
2. A
3. D
4. B
5. B
6. E
7. B
8. A
9. C
10. A

Chapter 2

The interrupted gene

General Concepts	Section	Related Questions
Introns/exons	2.1, 2.2	A1, A2, A3, A5, A7, D1, D2, D4, D6
Restriction endonucleases	2.3	A4, B4, D3
Gene organization and structure	2.4-2.8	A6, A10, A13, B3, C1, D4
Gene structure/protein function	2.10	A11, D5, D11
Gene families	2.11	A11
Regulatory networks and cascades	2.12	A15
Processes/structures		
Restriction mapping	2.3	B4
Gene isolation	2.6	A8
Alternative splicing	2.8	A11, B1, B2, C2, D12
Gene evolution	2.8	A10, A13, A14, B3, D4, D5, D7, D8, D9, D10
Overlapping genes	2.8	A12, B1
Examples, Applications and Techniques		
Electrophoresis	2.3	B4
Zoo blot	2.6	A8
Hybridization	2.6	A8, A9
Exon trapping	2.6	B5

Questions:

A. Recall and short answer

1. Sequences of genes that are represented in the mature RNA are called _____ whereas sections of genes that are removed from mature RNA are called _____.

2. The process that involves the removal of introns from RNA to give a messenger RNA is called _____.

3. Mutations in exons can cause changes in a protein's amino acid sequence whereas mutations in introns are more likely to affect _____.

4. A physical map of DNA can be obtained by breaking it at specific points defined by a target sequence of 4-6 base pairs. What is the name of the enzyme which allows such DNA breakage?

5. Introns are known to contain termination codons (UAA, UGA, or UAG) yet they don't interrupt the coding of a particular protein. Why?

17

6. Related genes tend to have a consistent organization with a similar number and position of introns. Variation in gene length in related genes is often attributed to variation in _____ length.

7. The mammalian gene DHFR (dihydrofolate reductase) is approximately 30 kb long, but the mRNA produced from that gene is only 2000 bases long. What accounts for the discrepancy between gene size and mRNA length?

8. A zoo blot is often employed to screen relatively short DNA fragments in several different species with the goal of identifying exons. Such a blot is dependent on labeled probes and _____ blotting.

9. Hybridization is a common technique used in molecular biology. What is meant by the term *hybridization*?

10. The number of introns varies among taxonomic groups. What general trends have emerged?

11. Homologous proteins that differ in the presence or absence of similar regions are generated by alternative splicing. How are such proteins generated?

12. Describe the process which gives *overlapping genes*.

13. Given that many genes in eukaryotes are interrupted, what two hypotheses have been presented to explain the origin of interrupted genes?

14. Provide an example of a gene which has undergone intron loss.

15. In eukaryotes, gene expression is often viewed in the form of molecular cascades. Provide an example of such a cascade.

B. Analytical and critical thinking

1. Homologous proteins can be formed by several different processes: overlapping genes and alternative splicing. Compare these two phenomena.

2. Present a general description of the role of alternative splicing among flies, worms, and humans.

3. One hypothesis suggests that exons function as mobile modules and act as building blocks in the evolution of genes. What evidence supports this hypothesis?

4. Assume that a piece of linear, double stranded DNA is digested and electrophoresed to give the fragments in the table below:

Restriction enzyme	Fragment size (bp)
A	600, 1000, 2100
B	1200, 2500
A + B	200, 600, 1000, 1900

What would be the restriction map of this piece of DNA?

5. Present an overview of the technique known as exon trapping. Why is such a procedure used?

C. Extensions and applications

1. Below are selected data from sequencing the genome of *Drosophila melanogaster* (data modified from Misra et al. 2002).

Total length of sequenced DNA	116.8 Mb
Total protein-coding genes	13,379
Protein-coding exons	54,934

(a) Assuming an average gene size of 5,000 base pairs, approximately how many bases are between genes?

(b) Approximately how many exons would one expect per gene on average?

2. (a) Researchers have shown that approximately 50% of human genes have alternative modes of expression (alternative splicing). In addition, approximately 15% of disease-causing mutations involve errors in alternative splicing (see Xu et al. *2002 Nucleic Acids Research* 30:3754-3766). What is meant by the term *alternative splicing*? How can mutations that disrupt alternative splicing lead to disease?

(b) Exon skipping appears to be a natural phenomenon to enhance dystrophin production in muscle cells of DMD (Duchenne muscular dystrophy) patients. The exon 45 deletion is the most frequent DMD-causing mutation while some individuals with a milder form of muscular dystrophy (Becker muscular dystrophy – BMD) have deletions in exons 45 and 46. Experimentally-induced skipping of exon 46 brings about a significant increase in dystrophin production (van Deutekom, J., and G. van Ommen, 2003 *Nature Reviews Genetics* 4:774-783). Provide a possible explanation for the enhanced production of dystrophin when a deletion or skipping of exon 46 occurs.

D. Standardized test format

1. What name is given to DNA sequences which are represented in the mature RNA in contrast to those which are removed when the primary transcript is processed to give the mature RNA?

    A. intron
    B. transcriptome
    C. exon
    D. split intron
    E. structural gene

2. When introns are spliced from RNA they are spliced only in *cis*, which means that

    A. splicing occurs in such a way as to form only *cis* enantiomers.
    B. all exons are spliced on the same molecule of the RNA.
    C. there are at least three different RNA molecules which rejoin as intermolecular hybrids.
    D. all introns are assembled from a variety of different RNA molecules to form a mature RNA.
    E. there are at least five different types of RNA which are involved in the assembly of introns into a single mature RNA.

3. Enzymes which cause specific breaks in double stranded DNA are called restriction endonucleases. Target sites of restriction endonucleases are often defined by

    A. chaperones.
    B. transcriptomes.
    C. multiple binding sites of snRNAs.
    D. specific base sequences of 4-6 base pairs.
    E. a multitude of protein-based binding factors.

4. Genes that are related by evolution have related organizations with most intron positions conserved. Variable lengths of genes are usually caused by variable lengths of

    A. exons.
    B. intergenic sequences.
    C. prions.
    D. poly(A) tails.
    E. introns.

5. If one compares the base sequences of related genes from different species one is likely to find that corresponding _____ are usually conserved but the sequences of _____ are much less well conserved.

    A. exons; introns
    B. introns; exons
    C. introns; chaperones
    D. chaperones; exons
    E. introns; proteins

6. The gene for Duchenne muscular dystrophy (DMD) is over 2000 kb but the mRNA produced by this gene is approximately 14 kb. What causes this discrepancy?

    A. The exons have been spliced out during mRNA processing.
    B. The DNA represented a double stranded structure, while the RNA is single stranded.
    C. There are more amino acids coded for by the DNA than the mRNA.
    D. The introns have been spliced out during mRNA processing.
    E. When the mRNA is produced, it is highly folded and therefore less long.

7. If one compares the frequency of uninterrupted genes and the overall size of genes among yeast, flies, and mammals, which cluster of terms is correct?

    A. yeast have the highest frequency of uninterrupted genes; mammals have the largest genes
    B. flies have the highest frequency of uninterrupted genes; mammals have the largest genes
    C. mammals have the highest frequency of uninterrupted genes; yeast have the largest genes
    D. yeast have the highest frequency of uninterrupted genes; flies have the largest genes
    E. flies have the highest frequency of uninterrupted genes; yeast have the largest genes

8. Various models have been proposed to explain the original form of genes that today are interrupted by sections of DNA called introns. The "introns early" model proposes that

    A. introns arose by gene duplication before operons and promoters evolved.
    B. the ancestral protein-coding units consisted of uninterrupted sequences of DNA into which introns were subsequently inserted.
    C. most of the genes of today are intronless because of alternative splicing.
    D. introns have always been part of genes and those without introns lost them in the course of evolution.
    E. introns evolved first in advanced eukaryotic genes and any gene without introns lost them through exon shuffling.

9. One fact suggesting that exons were the building blocks during gene evolution is that

    A. many exons can be equated with particular functions of protein sequences.
    B. many introns can be equated with particular functions of protein sequences.
    C. exons are generally larger than introns.
    D. introns are generally larger than exons.
    E. there are many more exons than introns.

10. All globin genes have a common form of organization with 3 exons and 2 introns. This suggests that

    A. all globin genes evolved independently from each other.
    B. their structure was formed by convergent evolution.
    C. their structure was derived from a multitude of ancestors.
    D. various forms of horizontal transfer drove convergent evolution.
    E. they are descended from a single ancestral gene.

11. In some cases there is a clear 1:1 relationship between exons and protein domains. An exception to this relationship could reasonably be explained by

    A. the removal of introns has fused adjacent exons without changing the integrity of coding.
    B. having more introns than exons in a gene.
    C. overlapping codons.
    D. having more exons than introns.
    E. allowing introns to be converted to exons.

12. Alternative splicing occurs when

    A. different combinations of introns are used in the production of a given mRNA.
    B. more than one intron is duplicated in a given mRNA.
    C. multiple introns are recovered in a given mRNA.
    D. a given exon undergoes duplication and forms two introns.
    E. different combinations of exons are used in the production of a given mRNA.

Answers:

A. Recall and short answer

1. exons, introns

2. RNA splicing

3. RNA processing

4. restriction endonuclease

5. Termination codons within an intron are not involved in translation because the introns are removed before translation.

6. intron

7. The 5 introns in the mammalian DHFR gene account for the discrepancy. Introns are spliced out of the RNA during processing.

8. Southern

9. Hybridization is the technique that assesses the formation of complementary hydrogen bonds between strands of nucleic acids. Usually a labeled probe is involved that identifies a complementary strand.

10. Introns are rare in prokaryotes, but common in eukaryotes. Among eukaryotes, introns appear in the following order of frequency (highest to lowest): mammals, insects, yeast.

11. Alternative splicing allows particular exons to be included or excluded.

12. Overlapping genes occur when two or more proteins are generated from a single gene by starting (or terminating) at different points during translation.

13. In the "introns early" hypothesis, genes are thought to have been interrupted early in their origin and those without introns have lost them during evolution. In the "introns late" hypothesis, the ancestral genes were intronless with introns being inserted subsequently.

14. The rat insulin gene has lost an intron. The ancestral gene had two introns, while one rat insulin gene now has one intron, the other has two introns.

15. During embryonic development, the expression of one set of genes leads to the expression of a second set, which in turn leads to additional sets of gene expression. Eventually, the orchestration of gene activity provides the adult organism.

B. Analytical and critical thinking

1. Overlapping genes occur when two proteins are generated from a single gene by starting (or terminating) at different points during translation. Alternative splicing involves the differential use of splice junctions. In both cases a given region of DNA leads to the production of related protein products but by entirely different mechanisms.

2. Alternative splicing broadens the number of proteins relative to the number of genes by about 15% in flies and worms. In humans, approximately 60% of the genes have alternative splicing modes of expression.

3. Gene structure is conserved among genes in very distant species suggesting that exon positioning is under selective pressure. Many exons can be related to protein sections that have a particular function. Related exons are found in different genes, suggesting a common origin for a given exon.

4. A restriction map of this section of DNA would be the following:

5. Exon trapping is a procedure which can be used to isolate exons from a pool of genomic fragments. A special vector is used which contains a strong promoter which drives the transcription of a single intron located between two exons. The intron contains a restriction site into which genomic fragments may be cloned. If an insert contains an exon with splice junctions, then splicing can occur and the RNA pools resulting will contain an RNA transcript of the exon. cDNA development by PCR would allow isolation of the exon.

C. Extensions and applications

1. (a) Assuming an average gene size of 5,000 base pairs there would be 5,000 X 13,379 equals about 6.7×10^7 base pairs comprising genes. Subtracting this value from 116.8 Mb gives 49.8 Mb between genes. Dividing 49.8 Mb by 13,379 gives about 3700 bases between genes. (b) 54,934/13,379 = 4.11 exons per gene

2. (a) Alternative splicing is the process which yields a variety of mRNAs when pre-mRNAs are spliced in more than one way. Various combinations of exons wind up in the final mRNA product. Precise removal of introns is required to maintain the coding integrity of exons. If, because of mutations, introns are not removed or not removed precisely, an aberrant protein product will be produced. Such abnormal proteins can lead to an altered phenotype. (b) It turns out that the deletion in exon 45 causes a reading frame shift leading to reduced dystrophin production. However, removal of exon 46 in conjunction with the exon 45 deletion, restores the reading frames and allows more dystrophin to be produced.

D. Standardized test format

1. C
2. B
3. D
4. E
5. A
6. D
7. A
8. D
9. A
10. E
11. A
12. E

Chapter 3

The content of the genome

General Concepts	Section	Related Questions
Levels of gene expression	3.1, 3.15, 3.16, 3.17	A1, A12, D1, D5
Linkage, restriction, and sequence maps	3.2	A2
Polymorphism	3.3	A2, D3
RFLP and SNP mapping	3.4	A3, D2, D3
Genome size	3.5	A5, A6, A7, B2, B3, C2, D6, D9, D10
Nonrepetitive and repetitive sequences	3.6, 3.12	A4, A11, B2, C2, D4
Gene numbers	3.7, 3.8, 3.11, 3.13, 3.14	A6, A12, B3, B6, C1, C2, D6
Types of genes	3.9	A8, B4, B5, D7, D8
Conservation of gene organization	3.10	A9, A10
Organelle DNA	3.18, 3.19, 3.20	A13, C1
Symbiosis	3.21	A13, B6, C1
Processes/structures		
Gene mapping	3.2	A2, A3, B1
Mapping to RFLP and SNP markers	3.4	A2, A3, D2
Mitochondrial DNA	3.20, 3.21	A13, B6, D11, D12
Chloroplast DNA	3.22	D11, D12
Examples, Applications and Techniques		
Chips and arrays	3.17	A2

Questions:

A. Recall and short answer

1. Three terms, *genome*, *transcriptome*, and *proteome*, are often used to describe the overall context of gene expression. Provide a brief definition of each.

2. Single nucleotide polymorphisms (SNPs) occur approximately every 1300 bases in the human genome. Of what use are SNPs in the study of human biology?

3. List three types of polymorphisms that are often used in gene mapping.

4. Distinguish between repetitive and nonrepetitive DNA. What are the relative proportions of each in representative organisms such as *E. coli* and the mouse?

5. Approximately how many genes are required to sustain a free-living organism?

6. The human genome contains approximately 30,000 to 40,000 genes. Is this more or less than the number originally expected?

7. Describe the origin and characteristics of a gene family.

8. Within gene families there are often copies of genes that have undergone sufficient alterations through mutations to become nonfunctional. Such nonfunctional entities are called_____.

9. When pairs of human and mouse chromosomes are compared for common gene location, many similarities are observed. This relationship is called _____.

10. Approximately what percentage of the mouse genes have a homologue in the human genome?

11. Repetitive sequences fall into five classes in eukaryotic genomes. List these classes.

12. How can the number of genes in a genome be less than the number of types of proteins in an organism?

13. Why do mutations in mitochondrial and chloroplast DNA lead to non-Mendelian inheritance patterns?

B. Analytical and critical thinking

1. The genetic map identifies the distance between mutations that affect the phenotype. Does the genetic map accurately reflect the physical map?

2. Evaluate the statement that there is no strong correlation between genome size and genetic complexity. Refer to the C-value paradox in your response.

3. Describe the relationship between genome size and gene number in bacteria and Archaea. Approximately how many base pairs on average are allocated to each gene? Does this relationship hold for eukaryotes? Explain fully.

4. What criteria are often used to identify an active gene within genomic sequence?

5. One method of determining the minimum number of genes necessary for free-living survival of an organism is to introduce random mutations or deletions and categorize ensuing phenotypes. When such an approach was used in yeast, results showed that of approximately 5,916 different mutated genes, only 18.7% were essential for growth on rich medium. Would you expect different results, in terms of the percentage of essential genes, if the experiment called for growth on minimal medium (medium lacking many nutrients)?

6. Modern free-living bacteria have at least 1500 genes. Mitochondria were thought to have originated when a primitive cell captured a bacterium that provided some benefit to the cell. Mitochondrial DNA contains as few as 13 genes. What process was probably involved in gene reduction in mitochondria?

C. Extensions and applications

1. The minimum number of genes needed to sustain a free-living lifestyle appears to be in the thermophile *Aquifex aeolicus* with a genome of 1.5 Mb and 1512 genes. In contrast, the bacterium *Borrelia burgdorferi*, the etiologic agent of Lyme disease, has a very dissimilar genome composed of a linear chromosome of about 0.91 Mb with 853 genes. There are also 17 linear and circular plasmids with a combined size of more than 0.53 Mb with approximately 430 genes (Frazer, C. et al. 1997 *Nature* 390:580-586). (a) Given that *Borrelia* is difficult to culture *in vitro*, and requires highly enriched medium and low oxygen tension, what would you suggest about its life style? (b) When successful, long-term *in vitro* culture of *Borrelia* causes changes in plasmid composition and protein expression profiles, as well as loss of infectivity. What function would you consider reasonable for the plasmids which inhabit *Borrelia*?

2. Analysis of multiplicons (groups of homologous chromosomal segments) of the *Arabidopsis* genome are presented in the table below (abbreviated from Simillion, C. et al. 2002 *PNAS* 99:13627-13632)

Chromosome no.	No. of genes in duplicated regions	Total no. of genes
1	5,532	6,488
2	3,163	4,023
3	4,335	5,096
4	3,027	3,738
5	4,637	5,832

(a) For each chromosome, what percentage of genes are formed in duplicated regions of the *Arabidopsis* genome?

(b) Is the frequency of genes in duplicated regions similar among different chromosomes?

(c) What percentage of the total number of genes in *Arabidopsis* is in duplicated regions?

(d) Given this degree of duplication and that the 2*n* chromosome number is 10, would you consider *Arabidopsis* to be diploid?

D. Standardized test format

1. The complete set of genes of an organism is called the _____ while the complete set of proteins of an organism is called the _____.

    A. genome, transcriptome
    B. variome, genome
    C. transcriptome, genome
    D. genome, proteome
    E. genome, variome

2. When comparing alleles sometimes there are single base changes. Such changes are called

    A. single protein polymorphisms.
    B. single nucleotide polymorphisms.
    C. products of alternative splicing.
    D. nuclear polymorphisms.
    E. none of the above.

3. Assume that two phylogenetically distinct genomes are compared by SNP and RFLP analysis. What trend is expected?

    A. The more alike the genomes the more variable the SNP and RFLP.
    B. If the two genomes are from closely related species, the SNP and RFLP will be more extensive than if the two genomes are from widely separated species.
    C. If the two genomes are from closely related species, the SNP and RFLP will be less extensive than if the two genomes are from widely separated species.
    D. There will be more SNPs than RFLPs because SNPs occur only in centromeric regions.
    E. The SNPs and RFLPs will be the same regardless of the phylogenetic relatedness of the two genomes.

4. Transposons of some eukaryotes may occupy more than half the genome. One could characterize a transposon in the following manner:

    A. short DNA sequences about 1 kb in length that have the ability to move in the genome
    B. relatively long DNA sequences >10 kb in length that have the ability to move in the genome
    C. relatively short DNA sequences about 1 kb in length that have stable positions in the genome
    D. moderately repetitive DNA segments that are found mainly around the centromere
    E. moderately repetitive DNA segments that are found mainly around the telomere

5. One plausible explanation for the observation of more types of proteins in a cell than apparent coding regions in the DNA is that

    A. mutations cause alterations in the translation process.
    B. there are many different promoter sites for a given gene in eukaryotes.
    C. alternative splicing allows for multiple proteins from a single transcript.
    D. conservation of intron/exon splicing junctions enable intron fusion.
    E. there are many cases in which different classes of introns constitute a common protein.

6. Approximately what percentage of the human genome consists of coding regions?

    A. about 1%
    B. about 20%
    C. about 40%
    D. about 80%
    E. 100%

7. Genes that code for corresponding proteins in different organisms are called

    A. homologs.
    B. comparalogs
    C. analogous.
    D. referenced genes.
    E. orthologs.

8. Generally, pseudogenes arise by

    A. gene duplications and subsequent mutations.
    B. neutral mutations.
    C. a variety of frameshift mutations.
    D. catastrophic deletions.
    E. catastrophic duplications.

9. The minimum number of genes required to sustain a free-living organism is approximately

    A. 50,000.
    B. 20,000.
    C. 10,000.
    D. 5,000.
    E. 1,500.

10. Assume that a prokaryote has a genome size of 10 Mb and contains 5,000 genes. Assuming that there are no non-coding regions in this prokaryote, what is the average gene size in terms of base pairs?

    A. 1,000 bp
    B. 200 bp
    C. 2,000 bp
    D. 5,000 bp
    E. 500 bp

11. The reason that mitochondria and chloroplasts can confer non-Mendelian inheritance patterns is because they

    A. contain DNA and are passed through the cytoplasm.
    B. duplicate genes in the nucleus.
    C. can be incorporated into nuclear genes.
    D. contain RNA as their genetic material.
    E. duplicate the genes in the nucleus.

12. In general, organelle genomes are

    A. circular DNAs which code for all the proteins found in the organelle.
    B. linear DNAs which code for all the proteins found in the organelle.
    C. circular RNAs which code for all the proteins found in the organelle.
    D. circular DNAs which code for some of the proteins found in the organelle.
    E. are highly consistent in their size and undergo extremely low frequencies of mutation.

Answers:

A. Recall and short answer

1. The genome is the complete set of genes of an organism. The transcriptome is the complete set of genes defined in terms of the RNA molecules produced. The proteome is the complete set of proteins from a genome.

2. SNPs illustrate the uniqueness of every human being and allow genes of interest to be mapped relative to SNP markers.

3. SNPs (single nucleotide polymorphisms), RFLPs (restriction fragment length polymorphisms), and minisatellites

4. Repetitive DNA consists of sequences that are present in multiple copies in a genome. Nonrepetitive DNA consists of sequences that are present in only one copy in a genome. Depending on the organism, nonrepetitive DNA may occupy most of the genome (prokaryotes) or a considerably smaller proportion (58%) in the mouse for example.

5. The smallest free-living bacterium, *Aquifex aeolicus* has a genome of 1.5 Mb and 1512 genes. *Mycoplasma gentalium* has the smallest known genome of about 470 genes but it is an obligate intracellular parasite.

6. The number of 30,000 to 40,000 is approximately 60,000 fewer than expected.

7. A gene family originates by gene duplication of an ancestral gene followed by the accumulations of mutations in the individual duplicated segments. Usually, members of a gene family have related but not identical functions.

8. pseudogenes

9. synteny

10. approximately 99%

11. transposons, pseudogenes, simple sequence repeats, segmental duplications, tandem repeats

12. The number of genes can be less than the number of proteins because alternative splicing allows for more than one protein to be produced by a single gene.

13. Because mitochondria and chloroplasts are cytoplasmic organelles and one sex, generally the female, provides more cytoplasm to the zygote than the other, the pattern of inheritance is nonchromosomal or non-Mendelian.

B. Analytical and critical thinking

1. The genetic map does not usually reflect the physical map of a chromosome because there are a number of factors which influence recombination frequencies relative to the physical locations of genes.

2. The C-value paradox relates to the apparent lack of a correlation between genome size and genetic complexity. For instance, even though we would consider humans as being more complex than toads, they have essentially the same genome size. In some phyla, there are large variations in DNA content among organisms that are morphologically quite similar. This is especially true in insects, amphibians, and plants.

3. For bacteria and Archaea, genome size is directly related to gene number and each gene occupies approximately 1000 bases on average. In eukaryotes, wide variation exists when comparing genome size with the number of genes. For example, in rice there are approximately 40,000 genes in a genome of 466 Mb, whereas in humans there are approximately the same number of genes in 3,300 Mb.

4. An active gene should consist of the following components:

   a. A series of exons should exist in which the first exon follows a promoter.
   b. The internal exons are flanked by appropriate splicing junctions.
   c. The last exon is followed by appropriate 3' processing signals.
   d. A single open reading frame should exist that starts with an initiation codon and ends with a termination codon. When the exons are spliced, the initiation and termination codons should be in frame.

5. Because a larger number of active genes would be required for growth on minimal medium (to make the nutrients not supplied), one would expect the percentage of essential genes to be significantly higher than 18.7%.

6. Mitochondria are not free-living and needed proteins and nutrients can be transferred from the cytoplasm to the mitochondrion. In addition, many of the genes originally in the mitochondrial genome have been transferred to the nuclear genome at a rate of approximately 2×10^{-5} per generation.

C. Extensions and applications

1. (a) *Borrelia* is more than likely not a free-living bacterium in that it requires extraordinary nutrients for survival. In addition, the requirement for low oxygen tension suggests that it resides within some other organism. (b) Because infectivity appears to be dependent on a particular plasmid population, it is likely that the plasmids encode proteins involved in the initiation and/or maintenance of infections.

2. Completing the table provides a response to portions (a), (b) and (c) in the problem.

Chromosome no.	No. of genes in duplicated regions	Total no. of genes	% of genes in duplicated regions
1	5,532	6,488	85.27
2	3,163	4,023	78.62
3	4,335	5,096	85.07
4	3,027	3,738	80.98
5	4,637	5,832	79.51
Total	20,694	25,177	82.19

(a) The percentages of genes in duplicated regions of the *Arabidopsis* genome are given above for each chromosome.

(b) The frequency of genes in duplicated regions is similar among different chromosomes.

(c) The percentage of the total number of genes in *Arabidopsis* is in duplicated regions is 82.19.

(d) It is clear that a relatively large percentage of the *Arabidopsis* genome is duplicated above the typical $2n$ level. Considering the above data, even though there are 10 chromosomes, the amount of intra- and interchromosomal duplications indicates a higher ploidy level. In fact, it has been suggested that there have been three large-scale duplication and polyploidization events in the history of this organism (Simillion, C. et al. 2002 *PNAS* 99:13627-13632).

D. Standardized test format

1. D
2. B
3. C
4. A
5. C
6. A
7. E
8. A
9. E
10. C
11. A
12. D

Chapter 4

Clusters and repeats

General Concepts	Section	Related Questions
Gene families and gene duplication	Intro. - 4.2	A1, A6, A8, A9, D9
Divergence of duplicated genes	4.2	A4, A9, D5
Gene loss and pseudogenes	4.2, 4.6	D1, B4
Neutral, silent and replacement variations	4.3, 4.	A2, A3, B1, D2
Conserved sequences, selection and evolution	4.4-4.6	A6, B3, C1, C2, D2, D7
Crossover fixation	4.10	B5
Processes/structures		
Unequal crossing-over	4.2, 4.7	C1, D6
Pseudogene formation	4.6	A2, A3, A6, B4, D1
Evolutionary clocks	4.4	A3, B1, B3, C2, D2, D7, D8
Heterochromatin and satellite DNA	4.11	A10
Examples, Applications and Techniques		
Globins	4.2-4.4	
LINES	4.5	
β-thalassemia	4.7	A7, D9
rRNA	4.8	A8, A9, D10
In situ hybridization	4.11	B6
Arthropod satellite DNA	4.12	B2, C1
Mammalian satellite DNA	4.13	A10, A11, D4, D11
Minisatellites as genetic markers	4.14	A11, A12, A13,C2, D12
Protein and DNA sequencing and content	all sections	B2, C2, D8

Questions:

A. Recall and short answer

1. Gene duplication appears to be a mechanism for the generation of new genes and gene families. What are two likely evolutionary consequences of a gene which has been recently duplicated?

2. What is meant by the term *functional gene* as contrasted with *nonfunctional gene*?

3. Categories of mutations accumulate at fairly constant rates after genes separate. However, _____ mutations accumulate more rapidly than _____ mutations.

4. By what mechanisms have all present-day globin genes descended from a single ancestral gene?

5. What is a pseudogene and how does one originate?

6. What types of observations would indicate that pseudogenes can persist in the genome over long evolutionary periods?

7. Thalassemias are blood-related diseases which result from loss of either the _____ or _____ chain of globin.

8. Provide a general description of rDNA transcription in an organism such as the newt, *Notopthalmus viridescens*.

9. The rDNA promoter of *Xenopus laevis* contains both repetitious regions and Bam islands. What are Bam islands?

10. Briefly describe the difference between heterochromatin and euchromatin.

11. What is meant by the term *hierarchical repeat*?

12. The acronyms VNTR (variable number tandem repeat) and STR (short tandem repeat) often refer to minisatellite and microsatellite DNAs. While not precisely defined, how many repeating bases make up each?

13. Why are VNTRs (variable number tandem repeats) and STRs (short tandem repeats) often used in DNA fingerprinting?

B. Analytical and critical thinking

1. Explain the differences among the following terms: *neutral mutation, silent mutation,* and *replacement mutation*. Are silent mutations more frequent than replacement mutations? Explain why, contrary to silent mutations, replacement mutations appear to accumulate linearly throughout evolution.

2. (a) Below is a predominant repeated sequence which comprises approximately 25% of the *Drosophila verilis* genome. Compute the buoyant density of the duplex DNA of this sequence.
          ACAAACT
          TGTTTGA

(b) Assume that the main band density of a sample of DNA is 1.704gm/cm^3. What would be the %GC in this sample?

3. Explain how one calculates divergence from an ancestral sequence. What do such calculations tell us about such divergence?

4. List and describe four mutational circumstances which may lead to the formation of a pseudogene.

5. Describe different outcomes from crossing over between tandemly-clustered genes. How are such outcomes different in nonrepetitive regions of DNA?

6. Provide an experimental protocol which would allow you to determine the actual physical location of satellite DNA in eukaryotic chromosomes.

C. Extensions and applications

1. One general feature of Arthropod satellite repeats is the presence of very long stretches of DNA with very low sequence complexity. Misalignment of such repetitive DNA strands followed by crossing over and DNA polymerase slippage are likely causes of variation of repeat number of *existing* repeats. However, many questions remain as to the initial source of such repetitive elements. Wilder and Hollocher (2001 *Mol. Biol. Evol.* 18:384-92) examined tetranucleotide microsatellites in several strains of *Drosophila* and observed a highly conserved short, nonrepetitive sequence just upstream (with no intervening bases) from the repeat in the following configuration:

$$5\text{'- (conserved, nonrepetitive)}(CPyGT)_n - 3\text{'}$$

For the purposes of this discussion, assume that such conserved elements are common in regions of highly repetitive sequences. What might be their significance?

2. Nuclear microsatellites have been used as markers for monitoring threatened populations because of their abundant polymorphism, wide dispersal, relative ease of access, and apparent selective neutrality. They are rarely associated with coding sequences of DNA. Variation in protein-coding sequences is determined by variations in electrophoretic mobility and the term "allozyme" is often used to describe a polymorph. Data below represent the percent polymorphism of allozymes and microsatellites in nuclear genomes of several groups of cats (Cheetah, Lion from Gir Forest, Puma from Big Cypress Swamp).

Polymorphism	*Cheetah*	*Lion*	*Puma*	*Domesticated Cat*
allozyme	4.1	0.0	4.9	21.3
microsatellite	84.1	19.3	42.9	98.9

(data adapted from Driscoll et al. 2002 *Genome Research* 12:414-423).

(a) Which cat group appears to have the greatest genetic variability as measured by allozyme polymorphisms?

(b) Which cat group appears to have the greatest genetic variability in terms of microsatellite polymorphisms?

(c) In general, which marker (allozyme or microsatellite) is most variable? Why?

(d) Which cat group would you consider to be in the most serious jeopardy of extinction?

D. Standardized test format

1. Pseudogenes have no coding function, but they can be recognized by sequence similarities with existing functional genes. How do pseudogenes arise?

    A. Pseudogenes arise by mutations within existing genes.
    B. They arise from simple sequence repeats (SSRs) in heterochromatic regions of the genome.
    C. They arise by the accumulation of mutations in formerly functional genes.
    D. Pseudogenes result from unequal crossing-over.
    E. They are the products of gene/promoter fusion.

2. Mutations accumulate about ten times faster at *silent* sites than at *replacement* sites. This is because

    A. silent mutations do not alter amino acid sequences of proteins.
    B. silent mutations occur in promoter regions only.
    C. there are more proteins which are silent.
    D. there are more genes that contain silent regions compared with replacement regions.
    E. silent mutations are not subject to repair.

3. A gene family refers to

    A. a set of genes that originate from the mother only.
    B. a set of genes that originate from the father only.
    C. a set of genes descended by duplication and variation from a common ancestral gene.
    D. any number of genes that can be grouped under the term *pseudogene*.
    E. a subset of genes that originate from the fusion of two parental genes.

4. When one centrifuges eukaryotic DNA on a density gradient, a separate peak is often formed which consists of a highly repetitive fraction of the genome. This peak is called_____ DNA.

    A. fractional
    B. subfractional
    C. density
    D. superficial
    E. satellite

5. Gene clusters are thought to be formed by

    A. gene duplication and divergence.
    B. gene deletion.
    C. gene inversion.
    D. gene repetition.
    E. none of the above.

6. Unequal crossing-over is a phenomenon which, based on recombination, leads to duplications and deficiencies in DNA segments. It is dependent on

    A. recombination between completely homologous sites.
    B. recombination between sites which are not homologous.
    C. gene slippage.
    D. DNA polymerase slippage.
    E. a combination of answers C and D above.

7. When one considers a nucleotide sequence of a coding region as either a replacement site or silent site, one is making a distinction between the likelihood that

    A. a mutation will or will not alter an amino acid in a given protein.
    B. a protein will lead to more than one amino acid change.
    C. the code is nonoverlapping or overlapping.
    D. there are fewer synonyms than coding sequences.
    E. there are three nonsense triplets for each amino acid.

8. In comparing syntenic (corresponding) sequences of mouse and human genomes, it has been determined that the proportion of sites that show signs of selection is about 5% whereas the proportion of sites that code for RNA or protein is about 1%. What conclusion is compatible with this information?

    A. The mouse contains more functional genes than the human.
    B. There are more functional genes in heterochromatin than was expected.
    C. Nonfunctional DNA suppresses coding DNA.
    D. There must be more functional DNA than nonfunctional DNA.
    E. There is a higher proportion of the non-coding DNA which is important from a
        selection standpoint compared to the coding DNA.

9. What is the general name for a human disease in which there is a reduction in the synthesis of either alpha or beta globin?

    A. alkaptonuria
    B. hemophilia
    C. hemochromotosis
    D. thalassemia
    E. albinism

10. The nucleolar organizer is a specialized region of the eukaryotic genome which is primarily involved in the synthesis of

    A. mitochondrial rRNA.
    B. ribosomal RNA.
    C. transfer RNA.
    D. protein-coding RNA.
    E. inhibitory RNA.

11. Satellite DNA is often found in _____ regions of chromosomes.

       A. euchromatic
       B. synthetic
       C. allelic
       D. chromatic
       E. heterochromatic

12. Which of the following is NOT an advantage of minisatellite and microsatellite sequences for gene mapping?

       A. abundant polymorphism
       B. constancy among species
       C. wide dispersal
       D. relative ease of access
       E. apparent selective neutrality

Answers:

A. Recall and short answer

1. The new gene may become a contributor to the survival of the organism and may be maintained by selection. This may occur if the new gene provides additional products of the same type (rRNA for example) or novel function (globins). If the new gene does not confer some selective advantage to the host, it is likely to be altered and inactivated by accumulated mutations.

2. A functional gene is defined as one which produces RNA and ultimately by the proteins for which they code. This definition may be broadened to include functional RNA species. Nonfunctional genes are defined by their inability to code for proteins usually caused by alterations in transcription, translation or both. Such nonfunctional genes are called pseudogenes and are symbolized as ψ.

3. silent, replacement

4. All present-day globin genes evolved from a single ancestral gene by a series of duplications, transpositions, and mutations.

5. A pseudogene is a DNA sequence that is similar to an existing functional gene but it has no coding function. They originate by the accumulation of mutations.

6. In some cases pseudogenes are duplicated along with functional genes and may share identical inactivating mutations and more sequence similarity than the functional genes to which they are related.

7. alpha; beta

8. In the newt, as in many other organisms including prokaryotes and eukaryotes, the major rRNAs are transcribed as a single precursor. Each transcription unit can be intensively transcribed such that many polymerases are closely packed and transcription products of increasing length are produced as the polymerases move down the rDNA.

9. Bam islands are regions of relatively constant sequence which separate the larger repetitious regions of the *X. laevis* promoter. They are called Bam islands because of their isolation by the BamHI restriction endonuclease.

10. Heterochromatin is the term which describes regions of chromosomes that are coiled tightly and relatively inert genetically. Euchromatin represents most of the genome and is usually found in the middle of the chromosome arms (excluding the centromeric and telomeric regions).

11. Hierarchical repeats are often found in mammalian satellite DNA where duplication and mutation of short repeating units leads to larger units which retain some characteristics of the shorter repeats.

12. The term microsatellite usually refers to repeating units less than 10 bp while minisatellite is used when referring to repeating units between 10 and 100 base pairs.

13. VNTRs (variable number tandem repeat) and STRs (short tandem repeat) are often used in DNA fingerprinting in eukaryotes because all such organisms possess such DNA and they vary greatly from individual to individual.

B. Analytical and critical thinking

1. Neutral mutations occur in genomic regions which do not code for a protein whereas both silent and replacement mutations occur within coding regions. Replacement mutations alter amino acid sequences whereas silent mutations only substitute one synonym for another so there is no change in the amino acid sequence. Because of the structure of the genetic code, replacement mutations occur in an approximate 3:1 ratio compared with silent mutations. However, when comparing DNA sequences, silent sites diverge at a greater rate than replacement sites. For some unknown reason, initial substitutions at silent sites are much more rapid than later substitutions. While such later substitutions are silent with respect to the protein, they likely come under some other form of selective pressure.

2. (a) Given the formula: $\rho = 1.66 + 0.00098(\%GC)\mathrm{gm/cm^3}$

the solution would be $\rho = 1.66 + 0.00098(28.57)\mathrm{gm/cm^3}$ $\qquad \rho = 1.688\mathrm{gm/cm^3}$

(b) Again, $\rho = 1.66 + 0.00098(\%GC)\mathrm{gm/cm^3}$, therefore,

$1.704\ \mathrm{gm/cm^3} = 1.66 + 0.00098(\%GC)$

$(1.704 - 1.66) = 0.00098(\%GC)$

$0.044/0.00098 = \%GC$

$44.9 = \%GC$

3. One identifies an ancestral sequence for a gene family by identifying the most common base at each position. The divergence of each member base at a given position is calculated as the proportion of bases which differ from the corresponding base in the ancestral sequence. Summations and averages of such divergence calculations provide an estimate of the degree of divergence among particular gene families.

4. The following mutational circumstances are known to cause the formation of pseudogenes:

   A. Promoter mutations: when a mutation occurs in a promoter, the gene cannot be effectively transcribed.
   B. Splice junction mutations: such mutations upset normal splicing processes such that pre-mRNAs may contain a mixture of intron and/or exon fragments.
   C. Nonsense mutations: when a termination codon occurs within a transcript, it leads to premature chain termination which usually interferes with protein function.
   D. Missense mutations: such mutations lead to amino acid substitution which may adversely influence protein function.

5. Crossing over in tandemly-repeated regions of the genome can generate variable lengths as a result of unequal recombination. In a region of nonrepetitive DNA, recombination occurs between two matched points on the homologous chromosomes because the unique sequences align precisely. Under this circumstance, expansion and contraction cannot occur.

6. Determination of the actual physical location of satellite DNAs could be determined by using *in situ* hybridization whereby labeled DNA is hybridized to fixed denatured chromosomes. Probe (labeled satellite DNA) detection allows one to directly visualize the location of the satellite DNA.

C. Extensions and applications

1. If it turns out that microsatellite islands are, in general, located in the area of a conserved sequence of bases, it is possible that those conserved sequences may provide the mechanism for the generation and/or maintenance of the microsatellite. It will be interesting to determine whether different microsatellite motifs have different conserved elements in their vicinity. On the other hand, it is possible that the microsatellite may generate the nonrepetitive region by as yet some unknown mechanism.

2. (a) The domesticated cat shows the greatest degree of allozyme polymorphism. (b) Again, the domesticated cat shows the greatest degree of microsatellite polymorphism. (c) Since allozymes are proteins, selection acts on them and mutations are less likely to be maintained in the population. Since selection appears to be less harsh on microsatellites because of their noncoding nature, they are likely to be more polymorphic. (d) Genetic variation is often considered to be a valuable survival resource. Generally, the more genetic variation, the greater likelihood that a population is genetically able to respond to environmental stress and habitat change. Since there is less genetic variation in lions (from the Gir Forest) they would appear to be the least genetically suited for long term survival. Perhaps through habitat destruction, predation, or disease, their numbers were reduced to a point which severely eroded their genetic variability. Domesticated cats on the other hand appear to be sufficiently variable so as to maintain their long term survival.

D. Standardized test format

1. C
2. A
3. C
4. E
5. A
6. B
7. A
8. E
9. D
10. B
11. E
12. B

Chapter 5

Messenger RNA

General Concepts	Section	Related Questions
Introduction	5.1	A1, D1, D2
Transcription and translation	5.2	B1, D3
tRNA forms a cloverleaf	5.3	A2, A3, B2, D4, D5, D6
Acceptor stem and anticodon are at opposite ends of tRNA	5.4	A4, D6
mRNA is translated by the ribosome	5.5	A5, B1, B4, D7
Many ribosomes bind to one mRNA	5.6	A6, D8
Life cycle of bacterial mRNA	5.7	B3, B6, D10, D11
Eukaryotic mRNA is modified	5.8	D10, D12
The 5' cap on eukaryotic mRNA	5.9	A7
3' polyadenylation of mRNA	5.10	A8, A10, D13
Multiple enzymes involved in bacterial mRNA degradation	5.11	A15
Determinants of mRNA stability	5.12	B5, C1, C2, D13
mRNA degradation requires multiple activities	5.13	A11, B7, D14
Nonsense-mediated mRNA decay	5.14	A12, A13, D15
Transport of eukaryotic mRNA	5.15	A14
Localization of eukaryotic mRNA	5.16	A15

Questions:

A. Recall and short answer

1. Why is RNA believed to have been the sole molecule responsible for the storage and expression of genetic information early in life's evolution?

2. In what sense does tRNA function as a molecular adapter?

3. What is the meaning of the following nomenclature: $tRNA_2^{Ser}$? What is the difference between $tRNA_2^{Ser}$ and $Ser\text{-}tRNA^{Ser}$?

4. All tRNAs have a similar overall shape, yet aminoacyl-tRNA synthetases can recognize their cognate (i.e the "right") tRNAs. In general terms, how is this possible?

5. The bacterial small ribosomal subunit is the 30S subunit and the large ribosomal subunit is the 50S subunit. What does S refer to? When the 30S and 50S subunits join, the complex is known as the 70S ribosome. What does this suggest about the relationship between units of S and molecular mass?

6. What is the difference between a ribosome and a polysome?

7. What is the function of guanylyl transferase in the creation of eukaryotic mRNA?

8. Explain the apparent paradox that virtually all eukaryotic mRNAs have a 3' poly(A) tail, yet there is no corresponding poly(T) sequence in the template strand of DNA.

9. How can the poly(A) tail of mRNA be used for purification of mRNA? Why is mRNA purified using this method enriched but not pure?

10. The 5' and 3' ends of eukaryotic mRNA interact. What bridges these two ends of mRNA?

11. What is the mechanism that connects deadenylation of eukaryotic mRNA and decapping of the same mRNA? What is the consequence of deadenylation and decapping?

12. Mutations are known that act as suppressors of nonsense-mediated mRNA degradation. What general function or functions do you predict for the wild type genes identified by these suppressor mutations?

13. Some mRNAs are transported to specific cellular locations. In this class of mRNAs, where is the determinant for transport located?

14. What serves as the "tracks" within a cell for mRNA transport? What is the "engine" for mRNA transport that runs along these tracks?

15. Explain the apparent paradox that the primary exonuclease that degrades bacterial mRNAs works in a 3'-5' direction, yet the overall direction of mRNA degradation is 5'-3'.

B. Analytical and critical thinking

1. How long would it take to transcribe a 100,000 nucleotide eukaryotic RNA? Assuming that rate of transcription is the same in bacteria and eukaryotes, and that the lengths of transcribed RNAs that are being compared are identical, which type of cell (bacterial or eukaryotic) would exhibit the shortest interval between the start of transcription and the first appearance of protein? Why?

2. If you wanted to alter a tRNA so it brought the incorrect amino acid to growing polypeptide chain (say inserting a histidine when the codon specified a cysteine), what would you alter?

3. In terms of gene expression, there is a distinct advantage to the instability of bacterial mRNA. What is this advantage? How would the bacterial cell's ability to respond to rapidly changing environments be different if its mRNA were to become stable?

4. Outline a procedure to determine how many amino acids of a growing polypeptide chain lie within the ribosome.

5. You are studying an mRNA that has an unusually short half-life and suspect it contains a destabilizing sequence element in its 3' UTR. Design an experiment to test this hypothesis.

6. Assuming no overlapping coding sequences, how many intercistronic regions would be present in a polycistronic mRNA that encodes 5 polypeptides?

7. Why are nonsense mutations in the terminal exon not subject to nonsense-mediated mRNA decay?

C. Extensions and applications

Some determinants of mRNA stability are in protein coding regions. This was suspected for the *c-fos* proto-oncogene mRNA. (Proto-oncogenes are genes that promote the development of cancer when they suffer a mutation that increases the activity of their product.) A major protein-coding region determinant of instability (mCRD) was hypothesized to exist in the 3' portion of the protein-coding region of the *c-fos* mRNA.

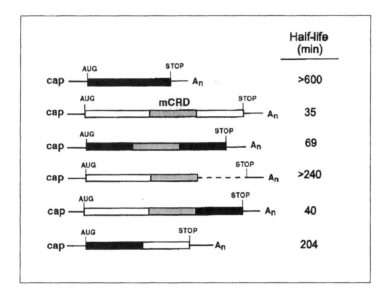

Grosset et al. (2000 *Cell* 103:29-40) performed experiments in which the suspected mCRD was inserted into the normally stable β-globin mRNA. The effects of these manipulations on mRNA half life were examined with the results shown above.

In this figure, β-globin mRNA coding sequences are shown by black boxes, the suspected *c-fos* mCRD is shown as a gray box, *c-fos* sequences not believed to be part of the mCRD are shown by open boxes, 5' and 3' UTRs are shown by lines, and the poly(A) tail is shown by A_n.

1. What general conclusions can be drawn from these results?

Additional experiments by Grosset et al. (2000 *Cell* 103:29-40) used selective deletion of the region between the mCRD and the poly(A) tail of the *c-fos* mRNA. The effects of deletion on mRNA half life were examined. The results are shown below.

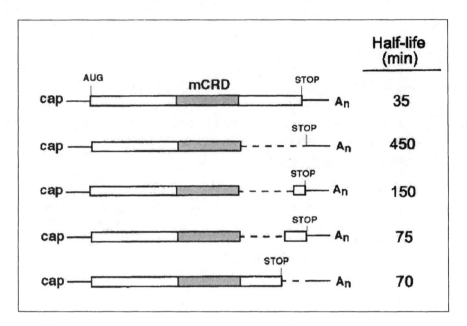

Solid lines indicate 5' and 3' UTR sequences, boxes show coding regions with the gray box being the mCRD, and A_n indicates the poly(A) tail.

2. What general conclusion or conclusions can be drawn from these results?

D. Standardized test format

1. RNA serves as a

    A. catalyst.
    B. information transfer molecule.
    C. structural molecule.
    D. A and B
    E. all the above

2. In mRNA, its _____ plays a dominant role in the function of the molecule, whereas its _____ is less important than it is for other classes of RNA.

    A. sequence; structure
    B. structure; sequence
    C. catalytic activity; sequence
    D. double-stranded nature; sequence
    E. ability to covalently link to amino acids; structure

3. A DNA sequence is: 5'AGGCTACTT3'
                                  3'TCCGATGAA5'

The mRNA sequence created by transcription of this sequence is: 5'AGGCUACUU3'

The upper DNA strand is the

    A. template strand.
    B. coding strand.
    C. antisense strand.
    D. plus strand.
    E. minus strand.

4. The enzymes that covalently link amino acids to tRNAs are

    A. tRNA acylases.
    B. tRNA polymerases.
    C. aminoacyl-tRNA synthetases.
    D. aminoacyl catalases.
    E. peptidyl transferases.

5. A(n) _____ bond links the _____ group of an amino acid to the _____ of a tRNA.

    A. ester; amino group; 2' OH of the 3' most residue
    B. peptide; amino; 3' OH of the 3' most residue
    C. ester; carboxyl; 2' or 3' OH of the 3' most residue
    D. ester; amino; 2' or 3' OH of the 3' most residue
    E. aminoacyl; carboxyl; 2' OH of the 5' most residue

6. The three-dimensional shape of a tRNA approximates

    A. a clover leaf.
    B. stem and loop.
    C. the letter L.
    D. the letter D.
    E. the letter O in the loop and the letter I in the stem.

7. A nascent polypeptide chain is a

    A. polypeptide chain in the process of synthesis.
    B. recently completed polypeptide chain.
    C. polypeptide chain not associated with the ribosome.
    D. protein that is part of the ribosome.
    E. protein produced by a polysome.

8. In a bacterium, ribosomes account for roughly _____ of the dry mass of the cell.

    A. 1/100
    B. 1/40
    C. 1/10
    D. 1/4
    E. 1/2

9. Polysomes function to

    A. increase the amount of mRNA that can be transcribed from a given gene.
    B. increase the pool of free ribosomes.
    C. increase the amount of protein that can be translated from a single mRNA.
    D. protect the mRNA from degradation beginning at its 5' and 3' ends.
    E. increase the rate of movement of a single ribosome along the mRNA.

10. Coupled transcription and translation does not occur in eukaryotes because

    A. of slower rates of transcription in eukaryotes relative to bacteria.
    B. of slower rates of translation in eukaryotes relative to bacteria.
    C. in eukaryotes, the rate of translation far exceeds the rate of transcription, but these rates are balanced in bacteria.
    D. in bacteria, the rate of translation far exceeds the rate of transcription, but these rates are balanced in eukaryotes.
    E. transcription and translation occur in one cellular compartment in bacteria, but in separate cellular compartments in eukaryotes.

11. The *E. coli lac* operon consists of three tightly-linked, co-transcribed genes. Therefore, mRNA transcribed from the *lac* operon is

    A. polycistronic.
    B. monocistronic.
    C. polysomal.
    D. monosomal.
    E. multifactorial.

12. Which of the following occurs before transcription of a eukaryotic gene is complete?

    A. Ribosome binding to the 5' end of the RNA
    B. The initiation of translation
    C. Addition of a 5' cap
    D. Addition of a poly(A) tail
    E. Nonsense-mediated mRNA decay

13. What is the function of the poly(A)-binding protein (PABP)?

    A. PABP stabilizes (i.e., blocks degradation of) mRNA at its 3' end.
    B. PABP stabilizes mRNA at its 5' end.
    C. PABP stabilizes mRNA at both its 5' and 3' ends.
    D. PABP prevents the premature initiation of translation.
    E. PABP prevents the premature termination of translation.

14. The exosome functions to

    A. degrade uncapped bacterial mRNA in a 5'-3' direction.
    B. degrade deadenylated bacterial mRNA in a 3'-5' direction.
    C. deadenylated yeast mRNAs in a 5'-3' direction.
    D. degrade uncapped yeast mRNA in a 5'-3' direction.
    E. degrade deadenylated yeast mRNA in a 3'-5' direction.

15. The RNA surveillance system acts to

    A. degrade mRNAs with nonsense mutations.
    B. degrade mRNAs with missense mutations.
    C. degrade all mRNAs with 5' UTRs.
    D. degrade all mRNAs with 3' UTRs.
    E. degrade all mRNAs with either a 5' or a 3' UTR.

Answers:

A. Recall and short answer

1.   RNA can serve as both an information storage molecule, catalyst, and structural molecule. Therefore, early in life's history, this "jack-of-all-trades" could have performed the current roles of DNA, RNA, and protein. In the context of genetics, RNA could store information, direct its own replication, transcription, and function as modern proteins do by serving catalytic, structural and regulatory roles.

2.   tRNA couples a nucleic acid code in the mRNA to amino acids that will be brought to the ribosome. In this sense it is an adapter between information in mRNA and amino acids that must be directed to the ribosome for protein synthesis.

3. $tRNA_2^{Ser}$ indicates the second tRNA to be named that can be charged with serine. $tRNA_2^{Ser}$ does not carry an amino acid. This is different from Ser-tRNA, because Ser-tRNA is an aminoacyl-tRNA, in this case serinyl tRNA, that is charged with serine. In the designation Ser-tRNA, the particular serinyl tRNA (1, 2 or any other) is not specified.

4.   Although all tRNAs are roughly the same size and shaped like the letter L, there are subtle structural differences that set them apart and allow a particular aminoacyl-tRNA synthetase to recognize its cognate (i.e., the "right" ones for the enzyme) tRNAs. Such structural determinants, which derive from the nucleotide sequence differences between tRNAs, include the pitch and length of double-stranded regions of the molecule and the conformation and length of the extra arm region.

5.   S is the abbreviation for a Svedberg unit, a measure of how rapidly a particle sediments during centrifugation. The fact that the assembled ribosome has an S value less than the sum of S values of the individual components hints at the fact that sedimentation velocity is not linearly related to mass.

6.   A ribosome is a single entity composed of a large and a small subunit and is devoted to translation. A polysome is the conglomerate of an mRNA being simultaneously translated by two or more ribosomes.

7.   Guanylyl transferase is the "capping enzyme" that adds a guanosine (G) residue through an unusual 5'-5' linkage to the 5' end of eukaryotic mRNAs. The G that is added is not methylated; methylation occurs after adding the G and is catalyzed by separate enzymes.

8.   The poly(A) tail is added post-transcriptionally (after transcription has occurred) to mRNA by poly(A) polymerase in the nucleus. Therefore, a run of T's in the DNA template strand that is complementary to the poly(A) tail is not found.

9.   The poly(A) tail can be used as a "handle" to pull the mRNA out of solution. This is done by allowing the mRNA to bind to short (usually ~15) strands of T residues (oligo-dT) attached to a matrix such as cellulose particles. RNAs without a run of As do not bind to the oligo-dT and can be washed away from the immobilized poly(A)-containing RNAs. The mRNA isn't perfectly pure because some classes of RNA other than mRNA also have extensive stretches of A residues that allow them to bind to the oligo-dT used to capture mRNA.

10. The two ends of the mRNA are bridged by the interaction of the poly(A)-binding protein (PABP) and initiation factor eIF4G that is in turn associated with the 5' cap of the mRNA.

11. When the poly(A) tail is shortened below a critical length (< ~10-15 nucleotides), PABP protein can no longer bind. Since PABP is a critical link between the 5' and 3' ends of mRNA, loss of PABP from the mRNA exposes the 5' cap, making it accessible to the decapping enzyme. Once mRNA is deadenylated and decapped, it becomes a substrate for nucleases that processively (nucleotide-by-nucleotide) degrade it from each end.

12. A mutation that suppresses nonsense-mediated mRNA decay blocks or reduces the efficiency of degradation of mRNAs that contain nonsense codons. These mutations would identify genes in the surveillance system to remove non-functional mRNA. These genes may encode proteins that are involved in any step of recognizing a ribosome that prematurely terminates, making the RNA a substrate for nucleases, or potentially in the nucleases themselves.

13. These localization determinants are almost always found in the 3' UTR region.

14. The "tracks" for mRNA transport are either microfilaments (actin filaments) or microtubules of the cytoskeleton. The "engines" are motor proteins such as myosin, kinesin or dynein that connect to a cargo (in this case, the mRNA) and use the energy of ATP hydrolysis to move along the cytoskeletal "tracks."

15. Although the direction of exonucleolytic (digesting a poly- or oligonucleotide from an end) digestion is 3'-5', the mRNA is broken into short fragments by an endonuclease (an enzyme that digests a nucleic acid at an internal residue) that makes cuts beginning near the 5' end and proceeding toward the 3' end. So in observing the overall direction of degradation, the bacterial mRNA becomes progressively shorter from its 5' end.

B. Analytical and critical thinking

1. The 100,000 nucleotide RNA would require roughly 2,500 seconds or 42 minutes for transcription (transcription occurs at a rate of ~ 40 nt/sec: therefore, 100,000/40 = 2,500). The first appearance of protein would occur much more rapidly in bacteria because there is no need for extensive processing of the primary transcript (the RNA produced directly from transcription) or for transport out of the nucleus.

2. By changing the nucleotides in the anticodon to recognize the codon for another amino acid, one could create a tRNA that would still be recognized by the same aminoacyl-tRNA synthetase but that would insert the incorrect amino acid into a polypeptide. For example, starting with a tRNA[his] and modifying the residues of the anticodon to recognize a cysteine codon, this altered tRNA would cause the insertion of histidine into positions normally occupied by cysteine in the polypeptide chain.

3. The advantage of an unstable mRNA is that the production of the encoded protein can be turned off rapidly when no longer needed. Shortly after transcription ceases, little mRNA remains. If bacterial mRNAs were stabilized, this rapid turn-off of protein synthesis would not be possible and cells would continue to synthesize proteins no longer necessary in an altered environment.

4.   In general terms, one could arrest translation, treat lightly with a protease to digest the nascent polypeptide chain exposed outside the ribosome (but not the ribosome), wash away or inactivate the nuclease, dissociate the assembled ribosome, mRNA, tRNA and the ribosome-protected nascent polypeptide, then run the protected nascent polypeptides on a gel to learn their average size.

5.   To learn if the mRNA contains a destabilizing sequence element in its 3' UTR, one could transfer the suspected element (by recombinant DNA technology) to the 3' UTR of a normally stable mRNA. If the stable mRNA is destabilized and if a control experiment that transfers an unrelated sequence to the stable mRNA has no effect on its stability, the conclusion would be that this sequence is an RNA destabilizing sequence element.

6.   There would be four intercistronic regions. The reasoning is that the number of intercistronic regions is equal to the number of cistrons (coding segments) –1. For example, for a polycistronic mRNA with two cistrons, there is one sequence between the genes, for a 3 cistron mRNA there are two sequences between genes, and so on.

7.   The junctions between exons are marked with a junction-binding protein or proteins. In most cases, the stop codon is contained in the last exon. If a nonsense mutation occurs in any exon other than the last one, it will cause arrest of the ribosome upstream of an exon junction. This will be detected as an aberrant (unusual) situation and will trigger mRNA decay. However, a nonsense mutation in the last exon could not be distinguished from a normal stop codon – both lie downstream of all exon junctions and neither would activate mRNA decay using a system that scans for ribosomes that terminate upstream of an exon junction.

C. Extensions and applications

1.   β-globin mRNA is stable and *c-fos* mRNA is much less stable. The suspected mCRD acts as a destabilization element. This is shown in the third line from the top of the figure. It appears that mCRD function requires some minimum length of mRNA between the element and the 3' end of the *c-fos* mRNA (see line 4). The mCRD is a distinct region of the *c-fos* mRNA rather than a general property of *c-fos* mRNA. This is shown in line 6 where a portion of *c-fos* mRNA not believed to contain the destabilization element does not significantly destabilize the β-globin /*c-fos* hybrid mRNA.

2.   mCRD function requires a minimum length of mRNA sequence between the destabilization element and the stop codon. From this set of experiments, it is unclear if this mRNA sequence must be a coding sequence or whether 3'UTR sequence would also function.

D. Standardized test format

1. E
2. A
3. B
4. C
5. C
6. C
7. A
8. D
9. C
10. E
11. A
12. C
13. C
14. E
15. A

Chapter 6

Protein synthesis

General Concepts	Section	Related Questions
Introduction	6.1	A1, A5, A12, B6, D1
Initiation, elongation, and termination	6.2	A2
Control of accuracy in synthesis	6.3	A3, B2, B3, D3
Initiation, the 30S subunit and accessory factors	6.4	A6, B1, D2
Initiator tRNA	6.5	A4, D5
IF-2 and f-Met tRNA	6.6	D5, D8
mRNA/rRNA pairing at initiation	6.7	B1, D6
Eukaryotic small subunit scanning	6.8	C1, C2, C3, D7
Complex of eukaryotic initiation factors	6.9	A6, A7, A8
Ef-Tu	6.10	A2, A14, B4, D8
Polypeptide chain transfer to aminoacyl -tRNA	6.11	A14, D12
Translocation	6.12	A10, D13
Elongation factors	6.13	A9, A14, B5, D9
Termination codons	6.14	D10
Release factors	6.15	A10, B5, D4, D10
Abundance of rRNA	6.16	A13
Active centers in ribosomes	6.17	A11, D14, D15
Role of 16S rRNA	6.18	A12
Peptidyl transferase activity of 23S rRNA	6.19	A12, A15, D12

Questions:

A. Recall and short answer

1. In general terms, how are bacterial ribosomes and non-organellar eukaryotic ribosomes similar? How are they different?

2. To what site on the ribosome does an incoming charged tRNA bind?

3. What stages of protein synthesis are subject to error? At which of these stages do errors occur most frequently?

4. In addition to AUG, the codons GUG and UUG are used in bacteria for translation initiation. How many different tRNAs recognize this set of initiation codons?

5. How many codons are covered by a translating bacterial ribosome?

6. What are major differences between the way translation is initiated in bacteria and eukarytotes?

7. How does the presence of the 3' poly(A) tail influence events at the 5' end of the mRNA critical for translation initiation?

8. In eukaryotes, what is the role of helicase activity in translation initiation?

9. How does diphtheria toxin disrupt translation?

10. What happens to make GTP-binding proteins dissociate from the ribosome?

11. Approximately how many amino acids of a nascent polypeptide chain lie in the exit channel of the eukaryotic ribosome?

12. What is the current view of the roles played by ribosomal proteins and RNA in ribosome function?

13. What does interaction between the 16S and 23S rRNAs accomplish during translation?

14. In general terms, how has investigation of the mechanism of action of antibiotics offered insights into ribosome function?

15. What evidence indicates that 23S rRNA harbors peptidyl transferase activity?

B. Analytical and critical thinking

1. You isolate suppressor mutations that restore the efficiency of translation to an mRNA with an altered Shine-Dalgarno sequence. What gene do you predict to be altered by these suppressor mutations? In what way do you think the gene will be altered? How do you think this suppressor mutation will affect the general health of the cell that carries it?

2. Imagine you have a collection of 1000 protein molecules, each 1000 amino acids long. Based on the range of error rates for protein synthesis *in vivo*, how many proteins in this collection do you expect to contain an incorrect amino acid?

3. Although the error rate in protein synthesis is impressively low, it is orders of magnitude greater than the error rate of DNA synthesis. Why can a much higher error rate in protein synthesis be tolerated relative to DNA synthesis?

4. What would happen if EF-Tu lost its ability to hydrolyze GTP (i.e., lost its GTPase activity)? Focusing on events occurring in the ribosome, would a mutation that deleted the EF-Ts gene lead to an identical outcome? Why or why not?

5. Just as RNA can serve in roles typically associated with proteins, proteins can mimic RNAs. Where in the process of translation do proteins mimic tRNA?

6. Many antibiotics target translation in either bacterial or eukarytotic cells, but not in both. How is it possible to achieve this specificity for a process that is central to both cell types?

C. Extensions and applications

Thompson et al. (2001 *PNAS* 98:12972-12977) investigated the ability of an internal ribosome entry site (IRES) to function in yeast. Previously, no IRES identified in mammalian or viral mRNAs had functioned in yeast. Thompson et al. worked with an IRES from cricket paralysis virus. To learn if this novel IRES functioned in yeast, they constructed a dicistronic (two coding sequence) mRNA with the viral IRES separating the first and second cistrons. This mRNA was added to an extract of yeast cells that supported translation and the production of each encoded protein was monitored.

The figure below shows the results of their experiment. In panel A, the protein-encoding segments of the mRNA are indicated by R-luc and F-luc, and the cricket paralysis virus IRES is shown by IGR-IRES. Panel B shows the wild type and mutant forms of the IRES, with underlined sequences indicating those that differ from the wild type. The N19 designation indicates a 19 nucleotide spacer between elements of the IRES sequence shown to be critical in insect cells. Panel C shows the results of *in vitro* translation of the dicistronic mRNA. These results are presented as the ratio of F-luc/R-luc. Δ indicates results obtained with a deletion of the IRES.

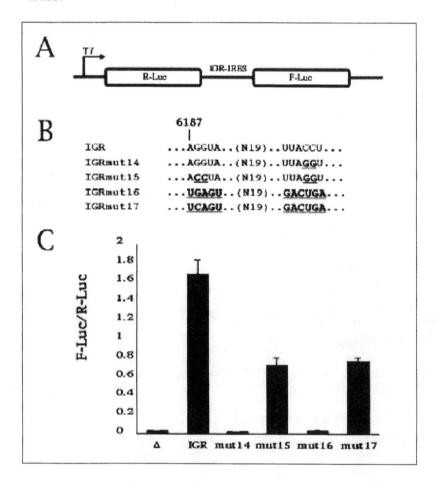

1. What is the evidence that the cricket paralysis virus IRES functions in yeast cell extracts?

2. Inspect the sequences shown for the wild type IRES. What appears to be a critical feature of the IRES?

3. Why do mutants 14 and 16 lack IRES activity and why do mutants 15 and 17 possess IRES activity?

D. Standardized test format

1. The major determinant of ribosome structure is

    A the proteins that contribute to the ribosome.
    B. the large RNAs that contribute to the ribosome.
    C. translation accessory factors.
    D . the tRNAs.
    E. the structure adopted by the mRNA.

2. How would an increase in the concentration of IF-3 affect the pool of non-translating ribosomes?

    A. Increased IF-3 would increase the proportion of free subunits relative to assembled ribosomes.
    B. Increased IF-3 would decrease the proportion of free subunits relative to assembled ribosomes.
    C. Increased IF-3 would decrease the proportion of ribosomes engaged in translation.
    D. Increased IF-3 would decrease the stability of ribosomes.
    E. Increased IF-3 would increase the stability of ribosomes.

3. The most frequent source of error in protein synthesis is

    A. incorporation of an incorrect base into RNA during transcription.
    B. binding of the incorrect tRNA to aminoacyl-tRNA synthetase.
    C. binding of the incorrect amino acid to aminoacyl-tRNA synthetase.
    D. binding of the incorrect charged tRNA to the codon.
    E. frameshift of the ribosome during translation.

4. The bacterial protein factor that is used for both initiation and termination is

    A. IF-1.
    B. IF-2.
    C. IF-3.
    D. RF1.
    E. RF2.

5. fMet-tRNA_f does not participate in the elongation phase of translation largely because

    A. it is not recognized by EF-Tu.
    B. it is not recognized by EF-Ts.
    C. it is not recognized by IF-2.
    D. it cannot recognize an AUG codon.
    E. it is too large to fit into the A site.

6. The Shine-Dalgarno sequence of a bacterial mRNA is recognized by

    A. a protein in the large subunit.
    B. a protein in the small subunit.
    C. a complementary sequence in the 5S rRNA.
    D. a complementary sequence in the 16S rRNA.
    E. a complementary sequence in the 23S rRNA.

7. In eukaryotes, the unit that scans down the mRNA to the first AUG codon is the

    A. 40S subunit.
    B. 48S complex.
    C. 60S subunit.
    D. 80S ribosome.
    E. 80S ribosome complexed with initiation factors.

8. Which of the following translation factors are most similar in their function?

    A. IF-2 and EF-Ts
    B. IF-2 and EF-Tu
    C. IF-3 and eIF-2
    D. IF-3 and RF1
    E. IF-3 and RF2

9. The structure of EF-G mimics that of

    A. an amino acid
    B. an aminoacyl-tRNA
    C. EF-Tu-GTP
    D. EF-Tu-GDP
    E. an aminoacyl-tRNA-EF-Tu-GTP ternary complex.

10. Bacterial class 1 release factors

    A. recognize stop codons.
    B. activate the ribosome to hydrolyze peptidyl-tRNA.
    C. separate the large and small ribosomal subunits.
    D. a and b.
    E. all the above.

11. tRNAs are large relative to the codon, yet adjacent codons in the P and A sites of the ribosome can be bound simultaneously by two tRNAs. This is possible because of

    A. a turn in the mRNA between the P and A sites.
    B. deformation of the tRNA anticodon loops.
    C. the ribosome's lowering of the free energy of codon/anticodon binding.
    D. the ribosome's increase of the free energy of codon/anticodon binding.
    E. the large size of the A and P sites relative to each tRNA.

12. The reaction responsible for peptide bond formation is

    A. amino peptidase.
    B. peptide synthetase.
    C. polypeptide polymerase.
    D. ribosomal translocase.
    E. peptidyl transferase.

13. The hybrid state model of ribosome translocation posits that

    A. the large subunit translocates first, the small subunit next.
    B. the small subunit translocates first, the large subunit next.
    C. a hybrid of large and small subunits translocates together.
    D. the assembled ribosome moves in two steps, the first one spanning two nucleotides, the second step spanning one nucleotide.
    E. the assembled ribosome moves in two steps, the first one spanning one nucleotide, the second step spanning two nucleotides.

14. The GTPase center of the ribososome

    A. has GTPase activity.
    B. loads GTP into the G protein.
    C. exchanges GTP for GDP on the G protein.
    D. exchanges GDP for GTP on the G protein.
    E. activates the GTPase activity of G proteins.

15. _____S rRNA interacts with the anticodon regions of tRNA in the A and P sites and 23S rRNA interacts with the _____ of peptidyl tRNAs in the A and P sites.

    A. 16; D loop
    B. 16; CCA terminus
    C. 16; codons
    D. 5; 16S rRNA
    E. 5; formyl methionine (f-met)

Answers:

A. Recall and short answer

1.   The general features of bacterial and eukaryotic ribosomes, including the presence of large and small subunits, the roles of RNA and proteins, the presence of closely-related rRNAs, and the possession of E, P, and E sites, are very similar. Ribosomes of these two groups of organisms differ primarily in the larger size of the eukaryotic ribosome.

2.   The A (aminoacyl) site.

3.   Every stage of protein synthesis is subject to error. These stages include the copying of the information in DNA into an mRNA that may change the meaning of the message, frameshifting of the ribosome during translation, binding of an incorrect aminoacyl-tRNA in the A site, and recognition by aminoacyl-tRNA synthetases of either an incorrect amino acid or an incorrect tRNA. Of these errors, the most common appears to be having an incorrect aminoacyl-tRNA occupy the A site (see Figure 6.8).

4.   Only one tRNA, fMet-tRNA, recognizes all three start codons in bacteria.

5.   Roughly 10 codons or 30 nucleotides of the mRNA are protected by the bacterial ribosome.

6.   In bacteria, the small subunit binds to the ribosome binding site in the mRNA. This occurs through interactions between the 16S rRNA and the ribosome binding site. In eukaryotes, there is no ribosome binding site. Instead, the small subunit associates with the 5' cap and then scans down the mRNA making use of the anticodon of the bound initiator tRNA to detect to the start codon. Almost always, this is the first methionine (AUG) codon encountered.

7.   The poly(A)-binding protein associated with the poly(A) tail recruits initiation factors that associate with the 5' end of the mRNA.

8.   Helicase first is required to unwind secondary structures near the 5' end of the mRNA to allow binding of other initiation factors. Later, it is needed to unwind secondary structures as the ribosome scans down the mRNA.

9.   Diphtheria toxin catalyzes the transfer of ADPR (adenosine diphosphate ribosyl) to eEF2 (the eukaryotic homolog of EF-G), thus inactivating it and shutting down translation.

10. When these proteins hydrolyze bound GTP to GDP, they change conformation and are released from the ribosome.

11. Roughly 50.

12. Proteins largely play a structural role in which they help maintain the catalytic rRNAs in optimal conformation.

13. This interaction holds the large and small subunits together.

14. Antibiotics that target translation have provided sensitive probes of ribosome function because of the great specificity of these agents. Information about normal ribosome function can be gleaned from learning exactly where the antibiotic binds and what it does to disrupt translation.

15. There are three main lines of evidence. 1) If ribosomal proteins are stripped from the large subunit while retaining the 23S rRNA, this preparation retains peptidyl transferase activity. 2) 23S rRNA prepared *in vitro* can catalyze peptide bond formation. 3) There is 23S rRNA in the region of the peptidyl transferase reaction active site, but no protein.

B. Analytical and critical thinking

1. Since the 16S rRNA interacts with the Shine-Dalgarno sequence, the most likely gene to be affected by this mutation is the 16S rRNA gene. The expectation is that the mutation creates a new sequence in the 3' region of the 16S rRNA. This altered 16S rRNA can now bind the altered Shine-Dalgarno sequence. This will interfere with normal cell function, as the mutation that reads the mutant Shine-Dalgarno sequence will not be able to recognize the normal Shine-Dalgarno sequence present on all other mRNAs.

2. Assuming an overall error rate of $1 - 10 \times 10^{-5}$ per amino acid incorporated, then the chance that any one protein contains an incorrect amino acid is $1 - 10 \times 10^{-5}$/amino acid x 1000 amino acids/protein = $0.01 - 0.1$ errors per protein. Since there are 1000 proteins in this set, then $10 - 100$ of them are expected to contain errors.

3. An occasional altered protein is unlikely to do significant harm. On the other hand, a similar error rate in DNA synthesis would wreak havoc on the cell since the DNA is an information storage device.

4. Without GTPase activity, EF-Tu could chaperone tRNA into the A site, but could not leave the A site. This block in the cycle of elongation would arrest translation. A deletion of the EF-Ts gene would yield an outcome different in detail but similar in overall effect. Since EF-Ts exchanges GTP for GDP on EF-Tu, loss of EF-Ts would prevent the regeneration of active EF-Tu. In this case, EF-Tu would be unavailable to chaperone aminoacyl-tRNAs to the A site. Translation would shut down (same overall result), but the details would be different. In this case, the A site would be unoccupied rather than occupied by an aminoacyl-tRNA-EF-Tu-GTP ternary complex.

5. In every step of elongation and termination, proteins mimic tRNA. EF-G, the complex of RF1/2-RF3, and RRF all mimic tRNA as it exists in a complex with EF-Tu.

6. Although the ribosomes of bacteria and eukaryotes are similar, they are not identical. Antibiotics that specifically act against translation in eukaryotes or bacteria exploit the structural differences between the ribosomes of these two groups of organisms.

C. Extensions and applications

1. In the case of the wild type IGR-IRES, abundant F-luc is produced (see bar labeled IGR in panel C). This is possible only if initiation is occurring at the internal site.

2. With one exception (the first U in the sequence at the right), these two sequences will base pair perfectly. This suggests that a double-stranded stem structure in the mRNA is important for IRES function.

3. These observations support the model deduced in the previous question. Mutations that disrupt base pairing in the IRES block internal ribosome entry. These are mutations 14 and 16. Compensatory mutations that restore a base paired structure (mutations 15 and 17) restore IRES activity, even if the sequence is different from the original.

D. Standardized test format

1. B
2. A
3. D
4. C
5. A
6. D
7. B
8. B
9. E
10. D
11. A
12. E
13. A
14. E
15. B

Chapter 7

Using the genetic code

General Concepts	Section	Related Questions
Introduction	7.1	A1, A2, B1
Wobbling and codon recognition	7.2	A3, A4, D1
tRNAs are processed	7.3	A5
tRNAs contain modified bases	7.4	A6, D3
Modified bases affect pairing	7.5	A7, D2
Alterations of the universal code	7.6	D5
Novel amino acids inserted at stop codons	7.7	A9, A10, D5
Synthetases and tRNA charging	7.8	A11, A12, D6
Two classes of synthetases	7.9	D7, D8
Proofreading by synthetases	7.10	A13, A15, B4, D7, D9
Suppressor tRNAs contain mutated anticodons	7.11	A8, D10, D14
Nonsense suppressors	7.12	B2
Suppressors compete with wild type reading of the code	7.13	A14, B2, D10
Ribosome enhances accuracy of translation	7.14	B3, D11, D12, D13
Recoding	7.15	A8, D10, D14
Frameshifting	7.16	B5, C1, C2, C3, D14
Bypassing	7.17	B5, D15
Active centers in ribosomes	7.17	

Questions:

A. Recall and short answer

1. What two experimental systems were employed to learn the meaning of codons?

2. Why might similar amino acids tend to be represented by similar codons?

3. Although there are 61 different codons that encode amino acids, any one organism has far fewer (generally around 40) different tRNAs. How is it possible to read 61 codons with less than 61 different tRNAs?

4. What are wobble rules?

5. tRNAs are synthesized as precursors that are processed to mature tRNAs. What are the processing steps?

6. Inosine is commonly found in the first position of the anticodon. What does this allow?

7. How is it possible for mutations outside the anticodon to change the codon pairing specificity (i.e., which codon it will pair with) of a tRNA?

8.   What type of gene must be mutated if a stop codon is to be read as a sense (amino acid-encoding) codon?

9.   What features in mRNA allow a UGA codon, normally read as a stop codon, to be read as a codon that specifies seleno-cysteine?

10. Seleno-cysteine and pyrrolysine are two non-standard amino acids that are specified by the genetic code. What is the process that links these unusual amino acids to tRNAs?

11. What are the two steps in aminoacylation of a tRNA?

12. How simple is it to predict which tRNA will be recognized by a particular aminoacyl-tRNA synthetase? What are some of the rules for how an aminoacyl-tRNA synthetase recognizes its cognate tRNAs?

13. What are the two steps at which an aminoacyl-tRNA synthetase can reject an incorrect amino acid?

14. Why is the UGA stop codon "leaky" so that it is read through in wild type cells at a frequency of 1-3%?

15. What are the direct recognition and kinetic proofreading models for how ribosomes increase the accuracy of translation? Which of these models appears to be correct?

B. Analytical and critical thinking

1.   The argument that all existing organisms are derived from a common ancestor is based in part on the near universality of the genetic code. Why isn't it argued that there were many ancestors of existing cells and that what we see today represents the outcome of convergent evolution in which many different early forms of genetic code evolved independently to the most efficient form – the one seen in life today.

2.   In *E. coli*, amber (UAG) suppressor mutations are easier to isolate and they suppress with higher efficiency than ochre (UAA) suppressors. Propose an explanation for this observation.

3.   Related but incorrect trinucleotides pair in solution with anticodons in tRNA at a frequency of roughly 1×10^{-2}. The frequency of mispairing aminoacyl-tRNAs to codons during translation is roughly 1×10^{-5}. How much does the ribosome improve the fidelity (accuracy) of codon/anticodon pairing?

4.   The error rate for substituting valine for isoleucine in rabbit gobin is $\sim 5 \times 10^{-4}$. The error rate of activating a tRNA^{Ile} with valine is $\sim 4 \times 10^{-3}$ and the rate at which a Val-tRNA^{Ile} (an isoleucine-accepting tRNA mischarged with valine) escapes the isoleucyl-tRNA synthetase also is $\sim 4 \times 10^{-3}$. What percentage of isoleucine $\rightarrow$ valine substitution errors in rabbit globin are due to errors that create Val-tRNA^{Ile}?

5. You are studying a retrovirus and believe that programmed frameshifting is occurring on one of its mRNAs. What type of product or products of translation would provide support for your model?

C. Extensions and applications

Barry and Miller (2002 *PNAS* 99:11133-11138) investigated the interaction between a "slippery sequence" frameshift element found in barley yellow dwarf virus and a downstream enhancer of this slippery sequence. The slippery sequence induces a –1 frameshift that is essential for the barley yellow dwarf virus to produce a DNA polymerase.

In an initial part of their study, Barry and Miller characterized the region around the slippery sequence. They did so by inserting the wild type and deleted forms of the slippery sequence-containing region between two genes that produced distinguishable and readily assayed products (reporter genes). They transfected (i.e., placed into) this construct into plant cells and assayed the production of the upstream protein (rLUC), which is produced without a frameshift, and the downstream protein (fLUC), which is produced only upon –1 frameshift. The results of this part of their study are shown in the figure below.

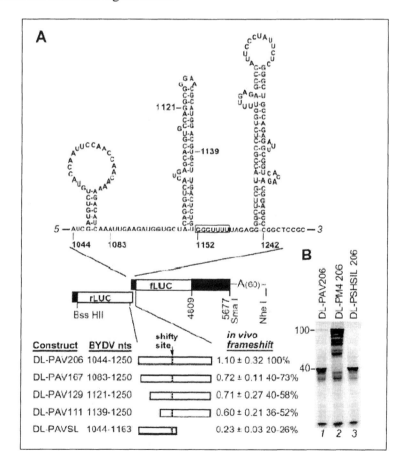

Potential secondary structures formed in this region are shown in the upper part of the figure with the slippery sequence boxed. The lower portion of the figure shows the wild type and deletion forms of the slippery sequence region inserted between the two reporter genes. The raw percentage and normalized efficiency of frameshifting is also presented. Numbering in the column "BYDV" indicates the nucleotides of the slippery sequence region that are present between the reporter genes. These positions also are shown in the sequence in the upper portion of the figure.

1. Based on this figure, what hypothesis was Barry and Miller testing?

2. Based on the results, what would you conclude?

3. There is one obvious deletion that is not shown in this figure. What is this deletion? There also is another construct that would be informative, but is not shown. What would this construct be?

D. Standardized test format

1. Mutations of the _____ of the anticodon often do not change the meaning of a codon.

    A. first base
    B. second base
    C. third base
    D. first and second bases
    E. second and third bases

2. Modifications of bases in the anticodon

    A. are not observed.
    B. disrupt codon/anticodon pairing.
    C. are an important determinant of tRNA specificity.
    D. are seen in all tRNAs.
    E. are restricted to the third base of the anticodon.

3. Base modification in tRNA occurs by

    A. incorporation of the modified bases during transcription of tRNA genes.
    B. spontaneous decay of unstable bases inserted into tRNA during transcription.
    C. synthesis of sequences of tRNA using a special RNA polymerase.
    D. enzymatic modification of standard bases in tRNA.
    E. pairing of tRNA with short RNAs that direct incorporation of modified bases.

4. Wobble rules predict that the minimum number of tRNAs required to ready 61 standard codons is

    A. 15.
    B. 22.
    C. 31.
    D. 42.
    E. 61.

5. Most alterations of the standard genetic code involve changing the meaning of

    A. sense (amino acid-encoding) codons to stop codons.
    B. one sense codon to that of another.
    C. a stop codon to a sense codon.
    D. a standard sense codon to a codon for a non-standard amino acid.
    E. a start codon to a stop codon.

6. Isoaccepting tRNAs are

    A. tRNAs with identical anticodons.
    B. tRNAs with an identical base in the second position of the anticodon.
    C. tRNAs with an identical base in the third position of the anticodon.
    D. tRNAs with an identical base in the second and third positions of the anticodon.
    E. tRNAs that will be charged with the same amino acid.

7. Aminoacyl-tRNA synthetases form critical contacts

    A. across the entire surface of a tRNA.
    B. along all residues on one face of a tRNA.
    C. with a limited number of residues concentrated at the extremities of the tRNA.
    D. only with the anticodon.
    E. primarily with the CCA residues at the 3' end of a tRNA.

8. One key difference between class I and class II aminoacyl-tRNA synthetases is

    A. class I and class II enzymes recognize opposite faces of the tRNA.
    B. class I enzymes recognize the anticodon whereas class II enzymes recognize the acceptor stem.
    C. class I enzymes recognize the acceptor stem whereas class II enzymes recognize the anticodon.
    D. class I enzymes contact the tRNA tightly across all residues of a single face whereas class II enzymes contact only residues at the extremities of the tRNA.
    E. class I enzymes contact only residues at the extremities of the tRNA whereas class II enzymes contact the tRNA tightly across all residues of a single face.

9. In kinetic proofreading by aminoacyl-tRNA synthetases, incorrect tRNAs

    A. associate slowly and dissociate rapidly.
    B. are aminoacylated slowly.
    C. are degraded if aminoacylation occurs.
    D. a and b.
    E. all the above.

10. A missense suppressor tRNA creates a mutator phenotype because the missense suppression

    A. is inefficient.
    B. is highly efficient.
    C. also leads to nonsense suppressor activity.
    D. will insert incorrect amino acids when reading wild type mRNAs.
    E. will lead to frameshifts when reading wild type mRNAs.

11. In solution (i.e., not on the ribosome), a correct codon will pair to the anticodon of a tRNA about _____ times as often as related but incorrect nucleotide triplets.

    A. 2-3
    B. 5-10
    C. 10 - 100
    D. 100 - 1000
    E. 10,000 - 100,000

12. 16S rRNA is believed to play an important role in increasing the fidelity of translation when

    A. the 3' end of the RNA pairs with and stabilizes correct codon/anticodon pairs.
    B. the 5' end of the RNA hydrolyzes the anticodon of incorrect codon/anticodon pairs.
    C. the central section of the RNA displaces an incorrect aminoacyl-tRNA.
    D. the 3' end of the RNA scans the tRNA acceptor stem to determine if the correct amino acid is attached.
    E. the 5' end of the RNA triggers the release of EF-Tu.

13. The delay between entry of an aminoacyl-tRNA into the A site and the peptidyl transferase reaction

    A. increases the accuracy of translation.
    B. decreases the number of prematurely terminated proteins.
    C. triggers premature termination.
    D. causes ribosomes to pause at stop codons.
    E. causes ribosomes to pause on "slippery" sequences.

14. A pause in translation can lead to frameshifting when

    A. a scarce aminoacyl-tRNA causes dissociation of the ribosome from mRNA.
    B. it triggers degradation of the mRNA
    C. the tRNA in the A site is released from the ribosome.
    D. the tRNA in the P site is released from the ribosome.
    E. the tRNA in the P site shifts along the mRNA.

15. The phenomenon of bypassing during translation

   A. occurs only in retroviruses.
   B. occurs in all transposable elements.
   C. is observed in many genes.
   D. is rare and occurs at virtually any sequence due to mistakes in translation.
   E. is rare and requires specific sequences and structures in mRNA.

Answers:

A. Recall and short answer

1. *In vitro* translation of synthetic RNAs and binding of trinucleotides of known sequence to aminoacyl-tRNAs

2. This reduces the negative effects of mutation. If similar codons encode similar amino acids, then single base mutations are more likely to lead to conservative substitutions of amino acids rather than contrasting amino acids. This is more likely to preserve protein function.

3. This is possible because of wobble pairing between the first base of the anticodon and the third base of the codon. These relaxed pairing rules allow one anticodon sequence to recognize more than one codon.

4. Wobble rules specify the possible non-standard base pairings between the first position of the anticodon and third position of the codon.

5. The steps of processing precursor tRNAs include trimming of 5' sequences with RNaseP and trimming 3' sequences with a combination of exo- and endonuclease activities, followed by CCA sequence addition to the 3' terminus of the tRNA, and finally modification of bases.

6. Inosine in the first position of the anticodon allows for very flexible pairing; C, U and A residues in the mRNA all can pair with inosine in the aniticodon.

7. These mutations alter the conformation of the anticodon loop, thereby altering the possible codons that it can pair with.

8. A tRNA-encoding gene must have its anticodon mutated so it recognizes the stop codon.

9. A stem-loop structure close to and downstream of the UGA stop codon must be present.

10. Standard amino acids are linked to tRNAs, then the amino acids are enzymatically modified while attached to the tRNA.

11. The first step is the formation of an aminoacyl~adenylate in which the amino acid is linked through its carboxyl group to the phosphate of adenosine monophosphate (AMP). In the second step, the amino acid is transferred to the 3' end of the tRNA with the accompanying release of AMP (see Figure 7.13).

12. Predicting the pairing between tRNA and aminoacyl-tRNA synthetase is not simple at all. The "rules" are loose and there are many idiosyncrasies. In general: 1) one base of the anticodon is recognized; 2) often one of the last three base pairs of the acceptor stem is recognized; and 3) there is an invariant "discriminator base" between the acceptor stem and CCA terminus characteristic of each set of isoaccepting tRNAs.

13. The aminoacyl~adenylate may be deadenylated, thus releasing the amino acid from the synthetase, and an incorrect aminoacyl-tRNA may be deacylated.

14. Although it does not follow the general rules of wobble pairing, the UGA codon is recognized inefficiently by the CCA anticodon of Trp-tRNA. For UGA stop codons, the wild type tRNA^{Trp} acts as an inefficient "natural" suppressor tRNA.

15. The direct recognition model postulates that the accuracy of translation is enhanced by the fit of the correctly paired aminoacyl-tRNA and anticodon within the ribosome's A site. The ribosome distinguishes between a correctly paired and an incorrectly paired codon/aminoacyl-tRNA pair and does not allow an incorrect pair to move forward to the peptidyl transferease reaction. In the kinetic proofreading model, the aminoacyl-tRNA must pass through two or more distinct steps prior to the peptidyl transferase reaction, with the correctness of the particular aminoacyl-tRNA assessed at each step. If an incorrect aminoacyl-tRNA is detected, it is released from the ribosome. In fact, proofreading appears to occur using elements of both models.

B. Analytical and critical thinking

1. There is no evidence that the current genetic code and its use in gene expression is the most efficient or the only one possible. It's likely that other systems could work as well. Therefore, the simplest explanation for why all life shares a common genetic core is that once an early cell evolved an efficient and highly interdependent system for gene expression, there was strong selective pressure not to modify any component as this would require compensatory mutations in many other components. The fact that all cells use essentially the same genetic system much more likely reflects the constraint against modifying the first efficient genetic system from which all current life is derived. Even if other genetic systems had evolved, the genetic system ancestral to our own allowed the line of organisms that carried it to out-compete others.

2. The amber stop codon is used less frequently in *E. coli* than the ochre stop codon. Since nonsense suppressors create C-terminal extensions of proteins translated from wild type mRNAs, there is a balance between the benefit of suppressing a nonsense mutation in an essential gene and creating problems by extension of the protein products of wild type genes. For this reason, the rarely used stop codon could be suppressed to higher efficiency as fewer normal mRNAs would be affected by read through of the stop codon.

3. About 1000-fold. Reasoning: the increase in fidelity is the ratio of the error rate in solution to the error rate during translation. This is $1 \times 10^{-2}/1 \times 10^{-5} = 1000$.

4. Roughly 4%. Reasoning: the mischarging error rate is the product of the errors at each step of the charging process. These steps are adding an amino acid to the tRNA and releasing the incorrectly charged aminoacyl-tRNA from the synthetase. Therefore, the mischarging error rate is $(4 \times 10^{-3}) \times (4 \times 10^{-3}) = 1.6 \times 10^{-5}$. The substitution error rate is 4×10^{-4}. Therefore, the percentage of errors contributed by mischarging tRNA^{Ile} with valine is: $(1.6 \times 10^{-5}/4 \times 10^{-4}) \times 100 = 4\%$.

5. You should look for a protein of the expected length and sequence predicted from the start codon to the standard stop codon, and in addition look for a larger protein that contains the protein produced by normal termination plus a C-terminal extension that runs from the first stop codon to a second stop codon that is +1 or −1 out of the reading frame of the normally terminated protein.

C. Extensions and applications

1.   They were investigating the hypothesis that the RNA secondary structure surrounding the slippery site is essential for frameshifting.

2.   Secondary structure of RNA surrounding the slippery site does play a role in frameshifting function. The effects of this set of deletions, however, appear relatively modest (~ 5-fold), with removal of the 3' secondary structural elements having the greatest effect.

3.   A key deletion that is not shown is deletion of the slippery sequence. Another construct that would have been informative is deletion of sequences on both sides of the slippery sequence. The presented deletions only involve secondary structures on one side or the other of the slippery sequence.

D. Standardized test format

1. A
2. C
3. D
4. C
5. C
6. E
7. C
8. A
9. D
10. D
11. C
12. A
13. A
14. E
15. E

Chapter 8

Protein localization

General Concepts	Section	Related Questions
Chaperones may be required for folding	8.4	D1, D3
Chaperones for newly synthesized and denatured proteins	8.5	A2, B1, D1, D2
Hsp 70 family is ubiquitous	8.6	A3, D2
Hsp60/GroEL forms an oligomeric ring	8.7	A4, B2, D2
Signal sequences initiate translation	8.8	A1, A5
Signal sequence interacts with SRP	8.9	A1, A5, D4
SRP interacts with SRP receptor	8.10	D5
The translocon forms a pore	8.11	D5
Translocon function	8.12	A6
Reverse translocation	8.13	A7, B3
Proteins reside in membranes via hydrophobic regions	8.14	D6
Anchor sequences determine orientation	8.15	D6
How do proteins insert into membranes?	8.16	B1
Post-translation insertion depends on leader sequences	8.17	A8, B5
Hierarchy of sequences determines location within organelles	8.18	A8, A9
Inner and outer mitochondrial proteins have different translocons	8.19	B4, D7
Peroxisomes employ another type of translocation system	8.20	B4, D8
Co- and post-translocation systems used by bacteria	8.21	A10
Sec system transports proteins into and through the inner membrane	8.22	B5, D9
Pores used for nuclear import and export	8.24	D10
Nuclear pores are large and symmetric	8.25	D11
Nuclear pore is a size-dependent sieve for smaller material	8.26	A11
Proteins require signals for transport through pore	8.27	B6
Transport receptors carry cargo protein through the pore	8.28	A13, C1, C3, D12
Ran controls direction of transport	8.29	A12, A13, C2, D12
RNA is exported by several systems	8.30	D13
Ubiquitin targets proteins for degradation	8.31	A14, D14
The proteosome degrades ubiquitinated proteins	8.32	A15, D15

Questions:

A. Recall and short answer

1. How do ribosomes anchored to the ER membrane differ from those that synthesize proteins in the cytosol?

2. Why are chaperones termed heat-shock proteins (Hsps)?

3. What role does ATP binding and hydrolysis play in the function of the Hsp 70 family of chaperones?

4. How does GroEL/GroES work in protein folding?

5. What are the characteristics of the signal sequence for proteins co-translationally inserted into the endoplasmic reticulum? What recognizes the signal sequence on nascent polypeptides?

6. How does the chaperone Bip act as a ratchet for proteins translocating to the lumen of the endoplasmic reticulum?

7. Why is there a need for reverse translocation from the ER to the cytosol? What proteins form the channel for reverse translocation?

8. There does not appear to be any sequence conservation for mitochondrial and chloroplast N-terminal leader sequences that direct the post-translational insertion of proteins into these organelles. Even so, these sequences are recognized by the cell and, with more difficulty, by researchers. What are the properties of these sequences and how can they function reliably if their primary sequences are not conserved?

9. Why do newly synthesized proteins bound for the lumen of chloroplasts or mitochondria contain two leader sequences?

10. What are four locations for proteins exported from the cytosol of bacteria?

11. Some proteins have an absolute requirement for a nuclear localization signal (NLS) for their localization in the nucleus, whereas other proteins present in the nucleus do not have a strict requirement for an NLS. What distinguishes these proteins?

12. How would inactivation of Ran-GEF alter transport into the nucleus?

13. What is the significance of the observation that a wide variety of nuclear import and export receptors compete with one another for binding to FG-nucleoporins?

14. When first discovered, ubiquitin was known only for its widespread (ubiquitous) distribution in all cells. What is the function of ubiquitin and why is it so ubiquitous?

15. Why is a significant fraction of newly synthesized proteins shunted to the proteosome?

B. Analytical and critical thinking

1.   Since chaperones recognize hydrophobic stretches of amino acids, why aren't the membrane-spanning regions of transmembrane proteins a constant site of chaperone action?

2.   It is possible to isolate mutant forms of GroES that cap GroEL but do not dissociate from it. Propose an experiment using this mutant form of GroES to test the hypothesis that multiple rounds of folding by GroEL/GroES are required for most proteins to acquire their final conformation.

3.   In a screen for mutations that inhibit protein degradation in the cytosol, mutations in components of the protein degradation system and in components of the translocator were isolated. How is it possible for translocator mutations to have an influence on cytosolic protein degradation?

4.   Chemical cross-linking of proteins normally translocated to mitochondria, chloroplasts, or peroxisomes prevents post-translational transport into the mitochondria and chloroplasts but has no effect on transport into peroxisomes. What is the reason for this difference?

5.   Mitochondria and chloroplasts both are derived from bacteria, yet when a comparison is made between the bacterial SecYEG translocator and eukaryotic translocators, homology is found between the Sec61 translocator of the ER and SecYEG, not between organelle and bacterial translocators. Why should this be?

6.   You hypothesize that a particular region of a protein acts as a nuclear localization signal. What features of this region might have promoted you to propose your hypothesis? How would you experimentally test this hypothesis?

C. Extensions and applications

The yeast protein Kap142p is an exportin for a variety of proteins. Yoshida and Blobel (2001 *J. Cell Biol.* 152: 729-740) examined the potential role of Kap142p as a nuclear transport protein for DNA replication protein A (RPA). RPA is a heterotrimer made of Rpa1p, Rpa2p and Rpa3 subunits and functions in the control of DNA synthesis.

In the experiment shown below, the location of an RPA fusion protein (Rpa2p fused to a tag termed PrA) in wild type yeast and a *KAP142* gene deletion strain (*KAP142Δ*) were compared. In the figure, DAPI is a stain for DNA that marks the nucleus and "Nomarski" is a light micrograph of the unstained yeast cells.

1. Based on this analysis, what conclusion should be drawn about the role of Kap141p in localizing RPA? Is this role consistent with Kap141p's previously assigned function?

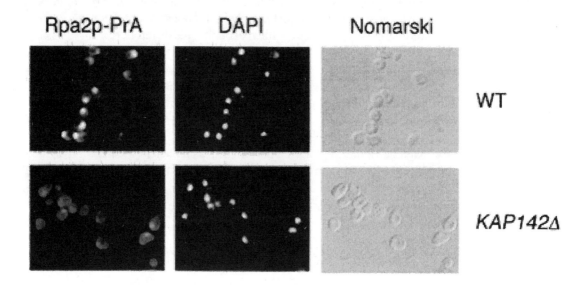

Cells of a yeast strain expressing the Kap142p-PrAp fusion protein were lysed and the fusion protein was attached to an affinity matrix that bound the PrAp tag. This procedure binds the Kap142p-PrAp fusion protein and any proteins that associate with it to the matrix. Next, the bound proteins were incubated in either buffer alone (lane 1), or in buffer with RanGDP (lane 2), or RanGTP (lane 3). After incubation, proteins released into the incubation buffer were collected, and run on a gel. Rpa1p was detected using an antibody. The results are shown below.

2.   Why is Rpa1p detected only when RanGTP is added? Is this consistent with the role of Kap142p in transport of RPA?

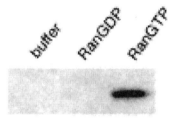

The role of Kap142 gene deletion on the localization of three other proteins was also examined. The results of this experiment are presented in the figure below.

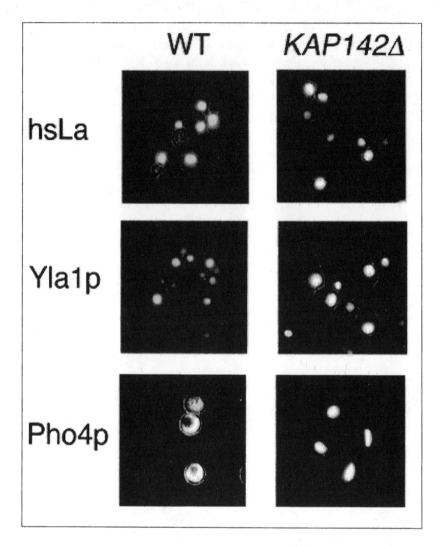

3. What conclusion can be drawn on Kar142p's effect on localization of each of these proteins? Is the effect the same on Pho4p as on RPA?

D. Standardized test format

1.   Chaperones bind to _____ of nascent, newly synthesized or denatured proteins.

      A. leader sequences
      B. amphipathic helices
      C. short basic or bipartite sequences
      D. tetrapeptide recognition sequences
      E. exposed hydrophobic regions

2.   The chaperone system that works after a protein is completely synthesized is the

      A. GrpE system.
      B. Hsp 40 system.
      C. Hsp 60/GroEL system.
      D. Hsp 70 system.
      E. Hsp 90 system.

3.   Which of the following is the best description of the way chaperones function?

      A. They prevent interactions with other proteins while allowing a protein to fold based on its amino acid sequence.
      B. They promote proper folding of proteins by providing a template for the misfolded protein.
      C. They use the energy of ATP hydrolysis to force the misfolded protein into the correct conformation.
      D. They change their conformation as GTP is bound and hydrolyzed, and as they do so, the interacting misfolded protein is brought into proper conformation.
      E. They chaperone misfolded proteins to the proteosome where these proteins are removed by degradation.

4. 7S RNA functions

      A. as the structural backbone of the SRP.
      B. to bind the SRP to the SRP receptor.
      C. to catalyze insertion of the signal sequence into the channel of the translocator.
      D. to bind the signal sequence.
      E. to stimulate GTPase activity of the SRP.

5. GTP hydrolysis to GDP by the SRP and SRP receptor

      A. allows the SRP to bind the SRP receptor.
      B. allows dissociation of SRP and SRP receptor.
      C. allows dissociation of the ribosome from the ER membrane after polypeptide synthesis.
      D. opens the seal on the cytosolic face of the translocator.
      E. opens the seal on the lumenal face of the translocator.

6. A signal-anchor sequence within a protein serves:

    A. to orient a transmembrane protein so its N terminus is in the ER lumen.
    B. to orient a transmembrane protein so its C terminus is in the ER lumen.
    C. to signal that a protein is to be anchored to the translocator.
    D. as a ratchet to prevent the escape of a protein from the translocator prior to completion of its synthesis.
    E. as a transmembrane anchor for co-translationally inserted ER proteins.

7. A transmembrane protein destined to reside in the inner mitochondrial membrane must

    A. pass only through the TOM translocator.
    B. first pass through the TOM translocator, then through a TIM translocator.
    C. pass only through a TIM translocator.
    D. first pass through a TIM translocator, then through the TOM translocator.
    E. first pass through a TIM translocator, then through the TOM translocator, then be inserted into the membrane.

8. Proteins are in their folded state when they are transported into the

    A. ER.
    B. mitochondrial intermembrane space.
    C. matrix (lumen) of the mitochondria.
    D. stroma of the chloroplast.
    E. peroxisome.

9. In translocation across the bacterial inner membrane, proteins are directed to the SecYEG translocon

    A. by binding of a signal sequence to the translocon.
    B. by binding of a signal sequence to the SecA protein that is associated with the translocon.
    C. by a chaperone that has an affinity for SecA.
    D. by a signal peptidase that has an affinity for the translocon.
    E. by 16S rRNA that binds the cargo protein and the translocon.

10. The rate of transport of proteins (of all types) into the nucleus through each nuclear pore is approximately _____ proteins per pore per minute.

    A. 2
    B. 20
    C. 75
    D. 350
    E. 2000

11. During transport into the nucleus, cytosolic proteins involved in associating cargo with the nuclear pore are _____ and those proteins of the nuclear pore complex required for translocation are _____.

    A. importins; nucleoporins
    B. nucleoporins; importins
    C. exportins; Ran
    D. Ran; importins
    E. NLSs; nucleoporins

12. Ran-GTP is present at higher concentration in the nucleus and Ran-GDP is at a higher concentration in the cytosol because of

    A. anchoring of Ran-GTP and Ran-GDP to proteins localized to the nucleus and cytosol, respectively.
    B. anchoring of Ran-GTP and Ran-GDP to different nucleoporins displayed on the nuclear or cytosolic faces of the nuclear pore complex.
    C. localization of Ran-GEF to the cytosol and Ran-GAP to the nucleus.
    D. localization of Ran-GAP to the cytosol and Ran-GEF to the nucleus.
    E. localization of Ran-GAP and Ran-GEF to the cytosol and their exclusion from the nucleus.

13. The major substrates for the nuclear export system are

    A. histones.
    B. proteins.
    C. RNAs.
    D. DNAs.
    E. ribonucleoproteins.

14. Ubiquitin ligase (E3 ligase)

    A. ligates ubiquitin to the ubiquitin-activating enzyme (E1).
    B. ligates ubiquitin to the ubiquitin-conjugating enzyme (E2).
    C. ligates ubiquitin to itself prior to transfer to target proteins.
    D. transfers ubiquitin from E2 to target proteins.
    E. ligates ubiquitin monomers into polyubiquitin complexes prior to polyubiquitin transfer to target proteins by the ubiquitin-conjugating enzyme (E2).

15. The 19S cap of the proteosome is responsible for

    A. adding ubiquitin to proteins.
    B. recyling ubiquitin chains stripped from degraded proteins.
    C. degrading ubiquitinated proteins.
    D. recognizing and transporting ubiquitinated proteins to the 20S proteosome.
    E. covering the open cylinder of the 20S complex to allow proteins to fold in isolation.

Answers:

A. Recall and short answer

1.   There is no intrinsic difference between ribosomes translating on the ER or free in the cytosol. The difference in location resides solely in the presence or absence of an N-terminal signal sequence of the nascent polypeptide produced by the translating ribosome.

2.   The name is historical and comes from the fact that proteins found to increase sharply in abundance after heat shock were logically named heat shock proteins. As it turns out, the majority of heat shock proteins are chaperones that come into play aiding the refolding of denatured proteins.

3.   ATP hydrolysis drives cycles of Hsp70 binding and dissociation from target proteins. The overall reaction is complex, involving initial binding of Hsp40 to an unfolded substrate, recruitment of Hsp70-ATP, assisted folding by the Hsp70/Hsp40 complex, ATP hydrolysis to ADP by Hsp70, displacement of ADP by GrpE, release of Hsp40, and finally release of Hsp70. The cycle starts anew when Hsp70 binds another ATP and recognizes an Hsp40 bound to a substrate protein.

4.   In general terms, GroEL provides a chamber for protein folding and GroES caps the chamber when a misfolded protein binds within the chamber. GroEL subunits bind ATP and this exposes hydrophobic residues within the wall of the chamber. These bind hydrophobic regions of misfolded protein. GroES recognizes a GroEL complex containing a protein and caps the open end of the chamber. This triggers ATP hydrolysis with the subsequent retraction of hydrophobic residues in the chamber wall. The protein folds in the now hydrophilic confines of the chamber. GroES uncaps the chamber and the protein that has gone through one cycle of folding exits. Typically, multiple cycles of entry, folding and release are required to complete the folding of a protein.

5.   The signal is composed of a run of 15-30 N-terminal amino acids that has a short stretch (~3-5) of polar amino acids followed by a longer run of hydrophobic amino acids. The signal sequence is recognized by the signal recognition particle, a ribonucleoprotein complex.

6.   As the polypeptide chain emerges from the lumenal pore of the translocator, Bip binds it. This acts as a ratchet by preventing the backward movement of the polypeptide chain out the translocator without inhibiting the forward movement driven by polypeptide synthesis and Brownian motion.

7.   Terminally misfolded proteins are degraded in the proteosome, a complex found exclusively in the cytosol. Since many ER proteins never fold properly, they must be transported out of the ER to the proteosome. The same translocator used for the insertion of proteins into the ER is used for their exit.

8.   In these leaders, particular amino acid sequences are not the key. Instead, the distribution of basic, polar, and hydrophobic amino acids, regardless of the unique identity of any amino acid, is the key. Mitochondrial and chloroplast targeting sequences both utilize stretches of uncharged amino acids separated by basic amino acids without any acidic amino acids present.

9. These organelles possess two membranes and proteins may be localized to either membrane, the intermembrane space or the lumen of the organelle. The first signal in the leader moves proteins through the outer membrane translocator and the second signal is required for movement through the inner membrane translocator.

10. Bacterial proteins moving from the cytosol may be destined for the inner membrane, intermembrane space (periplasm), outer membrane, or secretion from the cell.

11. Size is the key feature. Proteins less than 50 kD can move passively through the nuclear pore. These proteins do not have an absolute requirement for an NLS, although many of them possess an NLS as simple diffusion through the nuclear pore is slow. Proteins in excess of 50 kD are excluded from simple diffusion through the pore and must have an NLS to gain entry to the nucleus.

12. Transport into the nucleus would slow or shut down. Cargo is moved into the nucleus as a complex of cargo, an importin, and RanGDP. To dissociate the complex, GTP must be exchanged for GDP on Ran. This is the function of Ran-GEF, and without it, cargo would not be released and Ran could not be recycled to the cytosol to pick up additional cargo.

13. This suggests that exportins and importins all use a common mechanism of interaction with FG-nucleoporins to move through the pore.

14. Ubiquitin is covalently linked to proteins where it functions as a tag that marks these for degradation in the proteosome. Ubiquitin in so widespread because protein degradation is so pervasive; a significant fraction of the cell's protein forms components of the proteosome-based system of protein degradation.

15. This occurs because many newly synthesized proteins cannot fold properly. The high percentage of newly synthesized proteins that must be degraded indicates that the synthesis of a mature protein is an error-prone process.

B. Analytical and critical thinking

1. The hydrophobic regions of transmembrane proteins are never exposed to the cytosol, which is a signal for chaperone activity. Instead, the newly synthesized transmembrane region passes from the ribosome to the translocator pore then into the membrane.

2. If the mutant form of GroES is expressed to high levels, one could isolate the "trapped" GroES/GroEL complex, gently dissociate proteins of the complex so as not to denature individual proteins, and assay for the proportion of proteins that are fully folded. If the process of folding were processive (i.e., folding is completed before release of the protein), almost all GroEL/GroES substrate proteins would be folded. If the process required multiple rounds of entry and release from the GroES/GroEL complex, a large fraction of the proteins released from the GroES/GroEL complex would be incompletely folded.

3.   Translocator mutation can have an effect on cytosolic protein degradation because many proteins targeted to the ER do not properly fold and must be exported back to the cytosol where they are degraded in the proteosome. If translocator function is decreased, especially the reverse translocation required for movement of proteins from ER to cytosol, then there will be a decline in protein degradation within the cytosol.

4.   The difference stems from differences in the way proteins move through the organellar translocators. In the case of mitochondrial and chloroplast translocators, the protein must be unfolded. For the peroxisome, completely folded proteins pass through the translocator.

5.   The topology of translocation is the reason for the presence or absence of homology. In the case of secretion from the bacterial cell, ribosomes extrude a nascent polypeptide chain through a translocator across the membrane. This is essentially equivalent to the situation in which cytosolic ribosomes in a eukaryote extrude a protein across the ER membrane into the ER lumen. Therefore, the SecYEG and Sec61 translocators have a common function and origin. This is why they are homologous. The transport of proteins into organelles is equivalent to transporting a protein from the outside of a bacteria cell to its interior. This is a different process from inside-out movement and there is no reason to expect similar systems to perform these different tasks.

6.   You are likely to have noted a proline residue preceding a run of 3 – 5 basic amino acids. There are two experiments that could test the hypothesis. One test is to delete this sequence and see if the formerly nuclear protein becomes cytosolic. A second, more rigorous test is to transfer the putative NLS to a cytosolic protein and see if this fusion protein now localizes to the nucleus.

C. Extensions and applications

1. Kap142p is an importin for Rpa2p. This conclusion stems from the observation that in the absence of Kap142p (that is, in the *Kap142Δ* strain), Rpa2p-PrA is abundant in the cytosol. This is inconsistent with Kap142p's previously assigned role as an exportin. These observations do not suggest that previous results are incorrect; instead, they demonstrate that Kar142p acts as an exportin for some proteins and, at a minimum, as an importin for Rpa2p.

2. Rpa1p is detected only when the complex of Kap142p and associated proteins is incubated with RanGTP because RanGTP triggers dissociation of importin/cargo protein/Ran complexes in the nucleus. This is exactly what is expected if Kap142p functions as an importin for RPA.

3. Kap142p does not act as a nuclear transport protein (importin or exportin) for hsLa and Yla1p. However, it is an exportin for Pho4p. This conclusion comes from the fact that the distribution of hsLa and Yla1p is unaltered by deletion of the *Kap142* gene and the distribution of Pho4p, which is cytosolic in wild type cells, becomes nuclear in the deletion mutant. This is in contrast to Kar142p's role as an importin of RPA.

D. Standardized test format

1. E
2. C
3. A
4. A
5. B
6. B
7. B
8. E
9. C
10. D
11. A
12. D
13. E
14. D
15. D

Chapter 9

Transcription

General Concepts	Section	Related Questions
The transcription bubble	9.2	D1, D2
Three stages of transcription	9.3	A5, D5
Model for enzyme movement	9.5	A1, D4
Subunits of bacterial RNA polymerase	9.6	D3
Core enzyme and sigma factor	9.7	A2, A3, C3, D6, D7
Association with sigma changes at initiation	9.8	A4, A5, A6, D5, D6
A stalled polymerase can restart	9.9	D8, D10
How RNA polymerase finds promoters	9.10	A8
Sigma factor controls binding to DNA	9.11	A2, B1, D6, D7
Promoter consensus sequences	9.12	A4, D9
Promoter efficiency altered by mutation	9.13	A13
Polymerase binds to one face of DNA	9.14	B2, B3, C1, C2, C4, C5
Supercoiling is important in transcription	9.15	A10
Substitution of sigma factors may control initiation	9.16	A9, A11, A12, B4, D13
Sigma factors directly contact DNA	9.17	A7, A9, A13, B2, B4, D11
Cascades of sigma factors	9.18	A11, D13
Sporulation controlled by sigma factors	9.19	A11
Bacterial polymerase terminates at discrete sites	9.20	B14
Two types of terminators in *E. coli*	9.21	A14, D10
How does rho work?	9.22	B5, D12
Antitermination is a regulatory event	9.23	A15, D15
Antitermination sites	9.24	D14, D15
Termination and antitermination factors interact with RNA polymerase	9.25	B6, D15

Questions:

A. Recall and short answer

1. In outline, what path does DNA follow within a transcribing RNA polymerase?

2. What distinguishes the core enzyme and holoenzyme forms of *E. coli* RNA polymerase? What activity does the holoenzyme possess that is lacking in the core enzyme?

3. How does sigma factor association with the core enzyme change RNA polymerase's interaction with DNA?

4. What is the difference between a closed and an open complex of RNA polymerase and DNA? Which elements of the bacterial promoter are involved with closed and open complex formation? What is the evidence?

5. What marks the transition between initiation and elongation?

6. Once elongation commences, what happens to sigma factor?

7. Why does sigma factor only bind to promoters when complexed with the core enzyme?

8. What is the model for the way holoenzyme finds promoters?

9. A standard bacterial promoter contains –10 and –35 sequences that are highly conserved. The intervening nucleotides between these two promoter elements are highly variable in sequence, yet constitute a critical element of the promoter. How can this be?

10. Why are negatively supercoiled DNAs more readily transcribed than relaxed (non-supercoiled) DNAs? What are the roles of DNA gyrase and topoisomerase during transcription?

11. Why do bacteria have multiple sigma factors? Under what conditions are alternative sigma factors (sigma factors other than σ^{70}) deployed?

12. Promoters recognized by different sigma factors do not exhibit sequence conservation but they do exhibit conservation of other features. What are these conserved features?

13. Promoter mutations can be suppressed by mutations that affect the sigma factor. What does this result suggest? Why?

14. In a scan of the genome, what type of DNA sequence would you search for if the goal is to identify intrinsic terminators? Is this DNA sequence recognized by RNA polymerase to terminate transcription? If not, how is this sequence used?

15. How does antitermination constitute a form of regulation?

B. Analytical and critical thinking

1. When sigma factor associates with the core enzyme, it decreases the affinity of RNA polymerase (now the holoenzyme) for random sequence DNA by a factor of 10,000. What does this decrease in affinity accomplish? What would you predict to happen if such an affinity decrease did not occur?

2. In the footprinting technique, why is the DNA under investigation labeled at only one end?

3. Two approaches to examining the contacts made by RNA polymerase and the promoter are 1) to chemically modify bases in the promoter sequence prior to RNA polymerase binding in order to see which bases are critical for RNA polymerase recognition and 2) to allow RNA polymerase to bind an unmodified promoter, then see which bases are protected from chemical modification by polymerase binding. What is seen is that a greater number of bases are protected from chemical modification once polymerase is bound compared to the number of chemically modified bases that prevent polymerase from binding. What does this difference suggest?

4. Given the fact that the sequences of different classes of promoters are not conserved, why is overall promoter size and spacing of key elements conserved?

5. In a polycistronic operon (a set of genes cotranscribed into a single mRNA) a nonsense mutation in an upstream gene often blocks the production of all downstream gene products. This is known as a polar effect of the mutation. A missense mutation does not have this property. What explains polarity for nonsense mutations and lack of polarity for missense mutations?

6. The *nus* (N utilization substances) genes were isolated as mutations of *E. coli* that make pN of phage λ ineffective. What phenotype would you screen for in order to isolate *nus* mutations? What are the implications of the fact that *E. coli* mutations that suppress λ phage pN activity can be isolated at all?

C. Extensions and applications

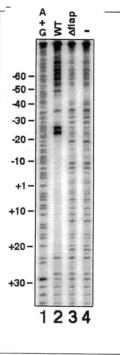

Kuznedelov et al. (2002 *Science* 295:855-857) investigated the role of the so-called flap domain of the β subunit of *E. coli* RNA polymerase. The flap domain extends outward like a pointing finger from the β subunit and was shown previously to play a role in converting a closed to an open transcription initiation complex. Kuznedelov et al. engineered a mutant form of the β chain that lacked only the flap domain. They reconstituted the holoenzyme with this mutant chain and using a standard footprint assay examined its interaction with a strong promoter. Their results are shown at right. Lane A + G is a marker lane (this shows where DNAs of known size run on the gel), WT is the holoenzyme reconstituted with the wild type chain, Δflap is the holoenzyme reconstituted with the flap-deleted form of β chain, and – is the same labeled DNA used in the WT and Δflap lanes but digested with DNaseI in the absence of any added protein. Numbers to the left are positions of nucleotides relative to the transcription startpoint.

1. What regions of the promoter show significant protection by the wild type holoenzyme?

2. What regions of the promoter show significant protection by the Δflap-containing holoeznyme?

3. What do you predict concerning the ability of the Δflap form of the holoenzyme to initiate transcription from this promoter?

A different promoter was used in a footprinting assay identical to the one above. The results are shown at right. This promoter has an extended –10 sequence and does not require the same contacts with holoenzyme as standard promoters.

4. What do you conclude about the effect of the Δflap mutation on holoenzyme binding to this promoter?

5. What do the arrow heads to the left of the figure indicate?

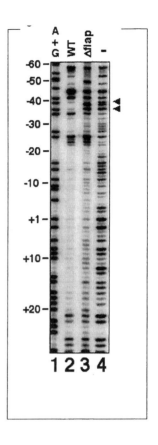

D. Standardized test format

1. The rate of transcription at 37°C in *E. coli* is roughly _____ nucleotides/sec.

    A. 5
    B. 10
    C. 40
    D. 200
    E. 800

2. The replication bubble is comprised of about _____ unwound base pairs of DNA in which the RNA-DNA hybrid extends for roughly _____ bases.

    A. 8 - 9; 3 - 4
    B. 12 - 14; 8 - 9
    C. 25 - 30; 18 - 20
    D. 50; 35
    E. 110; 72

3. The _____ and _____ subunits of the core enzyme surround the DNA and the _____ subunits act in enzyme assembly and interaction with activators of transcription.

   A. β; β'; α
   B. α; β; β'
   C. α; β'; β
   D. α; α'; β
   E. α; β; σ

4. The domain of RNA polymerase that maintains contact with its substrate as it moves down the DNA template is

   A. the wall domain.
   B. the bridge domain.
   C. the rudder domain.
   D. the contact domain.
   E. the β' domain.

5. Abortive initiation by RNA polymerase

   A. occurs only on mutant promoters.
   B. occurs only on weak promoters.
   C. occurs rarely on all promoters.
   D. occurs often on most promoters.
   E. occurs only when the transition between the closed and open complex is not made.

6. The core enzyme

   A. cannot bind DNA without sigma factor.
   B. binds DNA at any sequence to form unstable (half life < 1 min.) enzyme/DNA complexes.
   C. binds DNA at any sequence to form stable (half life > 60 min.) enzyme/DNA complexes.
   D. binds to promoters and awaits sigma factor association to initiate transcription.
   E. binds only the subset of promoters of genes that encode sigma and other accessory factors required for transcription initiation.

7. Within the cell, most RNA polymerase is

   A. bound to DNA.
   B. free (not bound to DNA).
   C. free and complexed with sigma factor.
   D. dissociated into separate subunits.
   E. bound to the membrane.

8. RNA polymerase also possesses RNase activity that comes into play when

    A. the polymerase stalls during elongation.
    B. the RNA portion of the RNA-DNA hybrid is cleaved at termination.
    C. the polymerase trims the 3' end of RNA back to the terminator sequence.
    D. the polymerase reads through a sequence that normally acts as a terminator.
    E. the polymerase converts a closed to an open complex.

9. A consensus sequence is one that is

    A. deduced from investigation of crystal structures of DNA binding proteins interacting with DNA.
    B. deduced from studying the effects of mutation on a single sequence, such as the promoter of one gene.
    C. deduced by comparing sequences, looking for perfect matches in all sequences.
    D. deduced by comparing sequences, looking for a high level of agreement, though not necessarily a perfect match, between all sequences.
    E. deduced from whole genome sequence analysis, looking for protein coding regions that are identical.

10. Pausing of RNA polymerase during elongation occurs

    A. at promoters.
    B. at regions rich in T's on the template strand.
    C. at regions of secondary structure in the newly synthesized RNA.
    D. at missense mutations.
    E. at stop or nonsense codons.

11. The secondary structure in sigma factor (and many other sequence-specific DNA binding proteins) that recognizes specific sequences in the promoter is

    A. an alpha helix.
    B. a beta sheet.
    C. a random coil.
    D. a molten globule.
    E. a hydrophobic pocket.

12. Rho functions by:

    A. promoting formation of hairpins in terminators.
    B. preventing formation of hairpins in terminators.
    C. enhancing the polymerase's ability to recognize hairpins in RNA.
    D. tracking along nascent RNA until it catches and displaces a paused RNA polymerase.
    E. cleaving the RNA/DNA hybrid at terminator sequences.

13. An important mechanism for coordinately regulating batteries of bacterial genes in response to an altered environment is by

    A. covalent modification of σ^{70}.
    B. production of alternative sigma factors.
    C. synthesis of antiterminator sequences.
    D. production of different core enzymes.
    E. creation of stress-response promoters.

14. Antiterminator sequences are located

    A. at the promoter.
    B. someplace between the promoter and terminator.
    C. adjacent to the 5' end of the terminator.
    D. within the terminator.
    E. adjacent to the 3' end of the terminator.

15. Antitermination works by

    A. creating an alternative hairpin structure in the RNA when the antiterminator is transcribed.
    B. suppression of the normal hairpin terminator by the antiterminator sequence.
    C. binding of antiterminator proteins at the <u>terminator</u> DNA sequence to prevent pausing by RNA polymerase.
    D. binding of antiterminator proteins at the <u>antiterminator</u> DNA sequence to prevent pausing by RNA polymerase.
    E. binding of antiterminator proteins to the RNA polymerase to modify its behavior at terminators.

Answers:

A. Recall and short answer

1. The DNA is bent with the transcription bubble immediately following the bend. Within the transcription bubble the single-stranded DNA "flips" bases outward to face the nucleotide entry site. DNA behind the transcription bubble is double-stranded while still in the channel of the RNA polymerase.

2. The core enzyme polypeptide composition is $\alpha_2\beta'\beta$. The holoenzyme is the core enzyme plus sigma factor. The holoenzyme can recognize promoters in the presence of sigma factor; the core enzyme alone cannot recognize promoters.

3. Without sigma factor, the core enzyme recognizes and binds to any sequence of DNA. Once sigma factor associates with the core enzyme, the holoenzyme has a greatly decreased affinity for random DNA sequences and acquires affinity for promoters.

4. A closed complex forms when the holoenzyme binds to a promoter prior to opening the two strands of DNA. The open complex is the next stage in which the holoenzyme opens the two DNA strands. The –35 sequence plays the predominant role in binding the holoenzyme to DNA. The evidence is that mutations of the –35 sequence reduce the probability of forming the closed complex but have little effect on the transition of closed to open complexes. The –10 sequence plays the dominant role in the transition from a close to an open complex. The evidence is that –10 sequence mutations have little effect on binding the holoenyzme to the promoter (closed complex formation) but slow the conversion of a closed to an open complex.

5. Initiation usually involves multiple abortive attempts to extend an RNA chain beyond the site where the open complex forms. Elongation begins only when the RNA polymerase begins to move down the DNA. This occurs when an RNA chain greater than ~ 9 nucleotides is synthesized.

6. In roughly 1/3 of cases, sigma factor dissociates as the polymerase begins elongation. Previously, it was believed that sigma factor dissociated from all elongating RNA polymerases. Current evidence indicates that in ~70% of cases sigma remains associated with the core enzyme as elongation begins. Sometime later the holoenzyme loses sigma factor.

7. As a free polypeptide, the amino-terminal domain of sigma factor blocks its own DNA binding sites (autoinhibition). When sigma factor binds the core enzyme, the autoinhibitory amino-terminal domain swings away from the DNA binding domains, unmasking their activity.

8. The current model posits that the holoenzyme binds DNA at random sequences. Once bound, it is rapidly exchanged to other DNA sequences until a promoter is encountered. The binding affinity for promoters is roughly 1×10^7 times higher than the affinity for random DNA, so once the holoenzyme is bound to a promoter, it forms a stable complex. It is important to realize that although DNA is a linear molecule, at the level of the genome, DNA is coiled and writhes, so a holoenzyme bound at one sequence will find itself in close physical proximity to many other sequences that are distant along the linear path of DNA.

9. Although the sequence of the spacer DNA is immaterial, its length is critical. If spacer length is altered, RNA polymerase cannot make contacts with the –10 and –35 sequences. Also, RNA polymerase makes critical contacts with some bases of the spacer, even though their sequence is irrelevant.

10. Negatively supercoiled DNA is underwound, having fewer twists of one strand about another relative to B form (the standard form) DNA. Negative supercoiling makes it easier to unwind DNA to form an open complex. During transcription, unwinding of strands by an advancing RNA polymerase will produce positive supercoiling (loosely, extra turns in DNA) ahead of the polymerase, and negative supercoiling behind the polymerase. DNA gyrase introduces negative supercoils into DNA. Gyrase acting ahead of the RNA polymerase counteracts the positive supercoils added to DNA. Topoisomerase relaxes negative supercoils. Working behind the advancing polymerase, topoisomerase prevents the DNA from acquiring excessive negative supercoils.

11. Alternative sigma factors are used for global (involving many genes) responses to new physiological conditions, especially stresses such as heat-shock, and starvation for carbon and nitrogen sources. These alternative factors are synthesized in response to the stresses, or in the case of σ^F, to the need to acquire motility. In some species, the stress of starvation unleashes a cascade of alternative sigma factors that triggers a developmental pathway resulting in the production of an environmentally-resistant spore.

12. These conserved features are the overall size and structure of the promoter and position of the transcription startpoint relative to the promoter. Each promoter has upstream and downstream sequence boxes (corresponding to the –10 and –35 sequences of σ^{70} promoters) of similar size and intervening DNA between these sequences of similar length.

13. This suggests that sigma factor directly contacts the promoter. This ability of sigma factor mutations to suppress promoter mutations is most simply explained by sigma directly contacting DNA. A change in sigma's amino acid sequence, especially in a recognition helix, would compensate for a change in promoter sequence.

14. You'd look for inverted repeats separated by a short sequence that is not part of the repeats. The inverted repeat would be followed closely by a sequence rich in A's in the template strand. For a strong candidate, the distal ends (the beginning of the left repeat and the end of the right repeat) of the inverted repeats should have a GC-rich sequence. This type of DNA sequence would allow the formation of a hairpin in the transcribed RNA followed by a U-rich, unpaired stretch. These are hallmarks of intrinsic terminators. The RNA polymerase does not recognize the sequence in DNA. Instead, it senses the formation of the hairpin in the newly transcribed RNA. This causes it to pause, increasing the probability of its dissociation, especially when a run of U's in the RNA is held within the transcription bubble.

15. Our understanding of regulation via antitermination has come mostly from the study of the regulation of bacteriophage lambda life cycle. Two clear examples are evident: 1)An antiterminator can allow RNA polymerase to read through a terminator within a gene (or between two regions of a gene), resulting in a longer product (See Fig. 9.51); 2) Some genes are separated from their promoter by a transcriptional terminator. At one stage of the life cycle, the gene is not to be transcribed, whereas at another, transcription of the gene downstream of a terminator is essential. This is where antitermination and its regulation come into play. Antitermination is activated only at specific antiterminators. This allows the polymerase to continue past the normal terminator into the downstream gene.

B. Analytical and critical thinking

1. The decreased affinity for random sequence DNA increases the ability of the holoenzyme to detect promoter sequences. Binding sigma factor creates an enzyme with an affinity for promoters, but if the high affinity for random sequence DNA was maintained, the holoenzyme's ability to distinguish promoter from random sequence DNA would be poor. Therefore, if the holoenzyme did not decrease its affinity for random sequence DNA, the efficiency of initiation would be reduced.

2. Since the technique requires visualization of a ladder of DNA fragments extending from one end of the molecule, labeling both ends would make interpretation impossible. Footprinting a DNA labeled at both ends would produce overlapping sets of fragments.

3. This suggests that a smaller number of bases are involved in recognition of the promoter by RNA polymerase than the bases that ultimately come to interact with the polymerase. This indicates that RNA polymerase makes initial contact with key bases of the promoter, then spreads to cover additional bases.

4. The sigma factors that recognize these promoters appear evolutionarily related. Since evolution is a conservative process, a straightforward way for sigma factors to diverge is by substitution of amino acids that change the types of DNA sequences that are recognized. If the overall structure of these diverged sigma factors is maintained, then conservation of promoter size and the spacing of its upstream and downstream sequence elements (elements homologous to the -10 and -35 sequences of σ^{70} promoters) as well as conservation of the length of the intervening sequence is an expected outcome.

5. Polarity is due to the relationship between rho and translating ribosomes on genes with rho-dependent terminators. Rho binds to the 5' end of mRNA and "chases" the RNA polymerase down the nascent RNA. Usually, ribosomes attach before rho and join in the chase of RNA polymerase. The translating ribosomes block rho from reaching the polymerase. If, however, the ribosome dissociates from the RNA, as it will at a stop codon, rho may catch the RNA polymerase. This is what normally happens at a standard stop codon. This also will occur at a nonsense codon. RNA polymerase pauses frequently at sites other than terminators during transcription. If rho is free to catch the polymerase as it transcribes RNA, it will. When this occurs, rho dissociates the polymerase from DNA, prematurely terminating transcription. In the case of a polycistronic operon, the premature termination occurs before the completion of downstream gene transcription. A missense mutation will not cause dissociation of ribosomes, and therefore not expose RNA polymerase to being overtaken by rho.

6. The screen would be for cells that do not produce (or produce greatly reduced levels of) phage λ proteins that require pN-mediated antitermination. It would be important to establish that the mutations were in the *E. coli* chromosome and not in the phage. The ability to isolate such mutations suggests that *E. coli* also uses a similar mechanism of antitermination to control expression of its genes (an interpretation supported by direct evidence). Without direct evidence that this is the case, an alternative interpretation that must be considered is that the Nus proteins influence pN-mediated antitermination, but have a different function in *E. coli*.

C. Extensions and applications

1.   The regions from roughly +10 to –18 and –27 to –35 show significant protection by the wild type holoenzyme.

2.   No regions show significant protection. In fact, the pattern for the Δflap mutant is indistinguishable from the control (-) run without any holoenzyme.

3.   It is very unlikely the Δflap mutant holoenzyme would support transcription initiation since there is no evidence the holoenzyme binds the promoter. (This prediction was borne out in subsequent experiments.)

4.   The Δflap mutant holoenzyme seems to have only a small effect on binding to this unusual promoter. (Subsequent experiments demonstrated that the Δflap mutant holoenzyme initiated transcription efficiently from this promoter.)

5.   The arrow heads show a difference in interaction between the wild type and Δflap mutant holoenzyme with this promoter. The wild type enzyme protects (and therefore appears to bind to) the –35 to –41 region of the promoter. This region is not protected by the mutant holoenzyme.

D. Standardized test format

1.   C
2.   B
3.   A
4.   B
5.   D
6.   C
7.   A
8.   A
9.   D
10.   C
11.   A
12.   D
13.   B
14.   B
15.   E

Chapter 10

The operon

General Concepts	Section	Related Questions
Introduction	10.1	D1, D2, D3
Regulation can be negative or positive	10.2	A2, C6, D4
Coordinate regulation of gene clusters	10.3	A1, C1, D3, D6
lac genes are controlled by a repressor	10.4	C1, C7, D5
The *lac* operon can be induced	10.5	A3, C6, C8, D3, D5, D7
Repressor is controlled by an inducer	10.6	A3, C4, C5, D8, D12
cis-acting mutations define the operator	10.7	A4, A5, A6, A9, B1, D2, D9
trans-acting mutations define the regulator gene	10.8	A6, A7, A9, B2
Special properties of multimeric proteins	10.9	A8, B2, D11
Repressor binds to the operator	10.10	A10, D13
Binding of inducer releases repressor from operator	10.11	A11, D14
Repressor monomer has several domains	10.12	A12
Repressor is made of two dimers	10.13	A12
DNA-binding is regulated by an allosteric change in conformation	10.14	A12, D8
Mutant phenotypes correlate with the domain structure	10.15	A7, D10
Repressor binds to three operators and interacts with RNA polymerase	10.16	A13, A14, C2, D15
Repressor is always bound to DNA	10.17	A15, B3
The operator competes with low-affinity sites to bind repressor	10.18	A15, B4, C3

Questions:

A. Recall and short answer

1. What advantage is offered to bacterial cells by transcribing polycistronic mRNAs?

2. When inducer is removed, *lac* mRNA levels decline sharply over the next few minutes but β–galactosidase levels remain steady over this period. What accounts for this difference?

3. The *lacY* gene codes for the β-galactoside permease required for transport of lactose into the cell. Since induction requires lactose within the cell, how can induction of the *lac* operon ever be triggered?

4. What is a *lacO^c* allele?

5. Explain why a mutation of a *cis*-acting site cannot be complemented by a wild type allele introduced into the cell.

6. What is the phenotype of a *lacI⁻* and *lacOᶜ* mutant? Using traditional genetic means (not sequencing the alleles), how could a *lacI⁻* mutation be distinguished from a *lacOᶜ* mutation?

7. Mutations of the *lacI* gene that affect the DNA-binding site and the inducer-binding site produce markedly different phenotypes. What are these phenotypes?

8. All *lacIˢ* alleles prevent inducer binding and some also act as dominant negative alleles. What would be the phenotype of a partial diploid carrying the *lacI⁺* and the dominant negative *lacIˢ* allele?

9. Predict the phenotype of a *lacOᶜ lacIˢ* double mutant.

10. Why is it not surprising that the operator is a palindrome?

11. One early model for how repressor interacts with inducer proposed that inducer binds to free (i.e., not bound to DNA) repressor, which then cannot bind to DNA. What makes this model implausible?

12. Which portion of the repressor undergoes the most dramatic conformational change upon inducer binding? How does this conformational change account for the repressor's function?

13. There are three operators (*O1*, *O2*, and *O3*) of the *lac* operon. *O1* is essential for repression. If either *O2* or *O3* are eliminated, there is a modest reduction in repression. However, if *O2* and *O3* are both deleted, repression is reduced roughly 100-fold. Explain these observations.

14. How is it possible for the repressor to enhance binding of RNA polymerase to the promoter while at the same time repressing transcription?

15. Assuming no change in the abundance of repressor protein, how would an increase in affinity of repressor for random sequence DNA affect regulation of the *lac* operon?

B. Analytical and critical thinking

1. In classic experiments, Jacob, Monod and their colleagues created partial diploids in *E. coli* in which they attempted to complement a variety of *lac* operon mutations. In the case of *lacOᶜ* mutations, introduction of the wild type allele to create a *lacOᶜ/ lacO⁺* partial diploid did not complement the original mutation. In contrast, introducing the wild type *lacI⁺* allele into a *lacI⁻* strain did result in complementation. Why the difference?

2. In later work, Jacob and Monod discovered that not all *lacI⁻* strains could be complemented by the wild type *lacI⁺* allele. Specifically, when a *lacI⁺/lacI⁻ᵈ* partial diploid was created, this strain continued to constitutively express the *lac* operon. Explain this result.

3. The equilibrium binding constant of RNA polymerase for the *lac* promoter is $2.5 \times 10^9 \text{ M}^{-1}$ when the operator is bound by repressor and $1.7 \times 10^7 \text{ M}^{-1}$ when the operator is not bound by repressor. Discounting competition from other promoters, in an *E. coli* cell (volume $\sim 1 \times 10^{-15}$ liter), what fraction of time will an RNA polymerase be bound to the *lac* promoter (length ~ 35 nucleotides) when repressor is bound to the operator? What fraction of time will an RNA polymerase be bound to the *lac* promoter when repressor is not bound to the operator?

4. The human genome size is roughly 3×10^9 bp. If a single operator was introduced into the human genome and one copy of repressor was expressed in a cell, what fraction of the time would the operator be occupied by repressor? (Assume all 3×10^9 bp are accessible to the repressor and that the affinity of repressor for the operator is 1×10^7-fold greater than for random sequence DNA.) What fraction of the time would the operator be occupied if 10 repressor proteins were expressed per cell? What fraction if there were 100 repressor proteins per cell?

C. Extensions and applications

Cronin et al. (2001 *Genes and Development* 15:1506-1517) attempted to create genetically engineered mice in which the investigator could regulate the expression of mouse genes using the *lac* repressor/operator system. After considerable effort, the investigators were able to induce the expression of model genes by adding IPTG to the mouse's water supply.

1. Although this is an interesting genetic trick, what uses could such a system be put to?

One of the gene constructs they prepared is shown in the figure below.

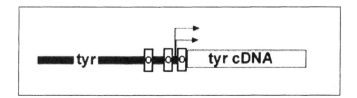

tyr indicates the tyrosinase gene. This gene encodes the enzyme tyrosinase which is required for the first step in melanin (the major pigment of the mouse) biosynthesis. Arrows mark the start sites of transcription. The black bar is the promoter region of the gene and the open boxes marked with "o" are *lac* operators.

2. Why are three operators present instead of one?

3. Given what you know about the way the repressor/operator complex interacts with *E. coli* RNA polymerase, if repressor can be expressed in mouse cells, is this construct almost certain to work or might it be a gamble?

Transgenic (genetically engineered) mice were created that had one copy of this construct integrated into the DNA of all cells.

4. For the tyrosinase gene to be regulated by IPTG, what other genetic construct had to be introduced into mouse cells?

Cronin et al. also created transgenic mice with the construct shown below.

SV40 is a strong mammalian virus promoter. The letters o and o_z show two forms of *lac* operator.

Cells from these mice were isolated and modified genetically so as to also express either the wild type (W) or one of three modified forms of the repressor. Levels of β–galactosidase activity were determined in each of these cell lines incubated in the presence or absence of IPTG added to the medium. The results are shown below.

Solid bars show the relative levels of β–galactosidase activity without added IPTG and stippled bars show levels with added IPTG.

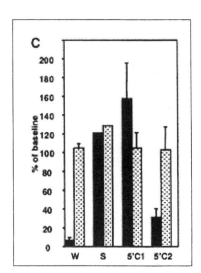

5. Which form of repressor (W, S, 5'C1 or 5'C2) is most effective for IPTG-regulated expression?

Beginning with an albino genetic background, transgenic mice were created that co-expressed the engineered tyrosinase gene and the most effective form of repressor. Through gestation and after birth IPTG was added to or excluded from the water given these mice and their mothers. The mice and close-ups of their eyes (which clearly show pigmentation) are presented in the figure below.

6. Why is the "Tyrosinase transgenic" mouse pigmented?

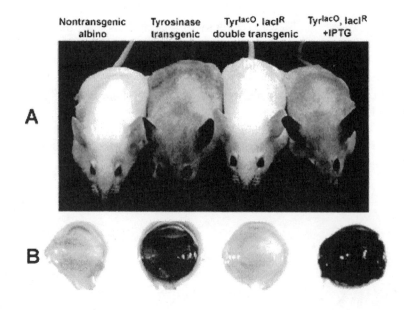

7. Why is the "Tyr^{lacO}, lacI^R" mouse white? (This mouse possesses the modified tyrosinase and *lac* repressor genes and was never exposed to IPTG.)

8. Why is the "Tyr^{lacO}, lacI^R + IPTG" mouse pigmented? (This mouse possesses the modified tyrosinase and *lac* repressor genes and the drinking water for it and its mother contained IPTG.)

D. Standardized test format

1. A reasonable working definition of a gene is:

   A. any sequence of DNA.
   B. a sequence of DNA that encodes a protein.
   C. a sequence of DNA that encodes a diffusible product.
   D. a sequence of DNA that contains one or more *cis*-acting regulators.
   E. a sequence of DNA that contains one or more *cis*-acting or *trans*-acting regulators.

2. Which of the following is an example of a *cis*-acting sequence?

   A. a promoter
   B. the operator
   C. the *lacI* gene
   D. a and b
   E. all the above

3. Which of the following is an example of a regulator gene?

   A. *lac*I
   B. *lac*Y
   C. *lacZ*
   D. a and b
   E. all the above.

4. In positive regulation

   A. the default state of the promoter is "off."
   B. the default state of the promoter is "on."
   C. a repressor must dissociate from the promoter.
   D. a repressor must bind an inducer.
   E. a transcription factor must bind to the repressor to block it from binding the operator.

5. Transcription of the *lacI* gene is controlled by

    A. the operator.
    B. the repressor.
    C. the repressor's interaction with the operator.
    D. a large set of transcription factors.
    E. the affinity of RNA polymerase for the *lacI* promoter.

6. A set of structural genes and the *cis*-acting elements needed to regulate them is

    A. an operon.
    B. a regulator.
    C. an operator.
    D. a polycistronic mRNA.
    E. an activator.

7. A corepressor is

    A. a small molecule that binds to and activates a repressor protein.
    B. a small molecule that binds to and inactivates a repressor protein.
    C. a protein that binds to and inactivates a repressor protein.
    D. a protein that binds to an operator to inhibit transcription.
    E. a DNA sequence that augments the effects of the operator.

8. Allosteric control in the *lac* operon is seen:

    A. when the repressor binds the promoter.
    B. when the repressor binds the operator.
    C. when inducer binding changes repressor shape and activity.
    D. when repressor inhibits RNA polymerase from initiating transcription.
    E. when repressor blocks RNA polymerase from binding to the promoter.

9. The fact that *lacO^c* operator mutations are *cis*-dominant predicts that

    A. a mutant operator introduced into a wild type cell will prevent transcription of the *lac* operon.
    B. a mutant operator introduced into a wild type cell will result in constitutive transcription of the *lac* operon.
    C. a wild type operator introduced into a cell with a *lacO^c* mutant operator will rescue the mutant phenotype.
    D. a wild type operator introduced into a cell with a *lacO^c* mutant operator will fail to rescue the mutant phenotype.
    E. a wild type operator introduced into a wild type cell will enhance operator activity.

10.   The repressor produced in *lacI^s* mutants

   A. binds the promoter constitutively.
   B. binds the promoter when repressor binds inducer.
   C. binds the operator constitutively.
   D. binds the inducer constitutively.
   E. cannot repress transcription initiation.

11.   The *lacI^d* allele is a dominant negative allele. This implies that

   A. in a strain carrying only the *lacI^{-d}* allele, the mutation is dominant.
   B. in a strain carrying only the *lacI^+* allele, the wild type allele is recessive.
   C. a strain carrying both the *lacI^{-d}* and the *lacI^+* alleles will produce homotetramers
   formed solely of mutant repressor and homotetramers made solely of wild type repressor.
   D. a strain carrying both the *lacI^{-d}* and the *lacI^+* alleles will produce repressor tetramers
   that are mixtures of mutant and wild type repressor polypeptides.
   E. a strain carrying both the *lacI^{-d}* and the *lacI^+* alleles will be inviable.

12.   In experiments that investigate how inducer regulates repressor, IPTG is used instead of the
physiological inducer because

   A. IPTG binds repressor with nearly 100-fold greater affinity.
   B. IPTG binds repressor with nearly 100-fold less affinity.
   C. IPTG freely crosses the cell membrane.
   D. IPTG is transported much more efficiently by β–galactoside permease.
   E. IPTG cannot be hydrolyzed by β–galactosidase.

13.   To say the operator is a palindrome is to say it is

   A. a double-stranded sequence.
   B. a double-stranded sequence in which only one strand is important for repressor
      recognition.
   C. a sequence with dyad symmetry ($180°$ rotation gives the same sequence).
   D. the site of repressor binding.
   E. a sequence that overlaps the promoter.

14.   Binding of inducer to repressor

   A. changes the orientation of the DNA-binding headpieces relative to inducer-binding
   core.
   B. changes the orientation of the DNA-binding core relative to the inducer-binding
   headpieces.
   C. causes the repressor to separate into two dimers.
   D. causes the repressor to separate into four monomers.
   E. causes the DNA-binding dimers of the repressor to associate into a tetramer that lacks
   DNA-binding activity.

15.   When bound to the operator, repressor

    A.   completely blocks RNA polymerase from binding at the promoter.
    B.   reduces the affinity of RNA polymerase for the promoter.
    C.   excludes RNA polymerase from binding to any DNA sequences within ~ 500 nucleotides of the promoter.
    D.   increases the affinity of RNA polymerase for the promoter.
    E.   tethers RNA polymerase to itself (the repressor) so RNA polymerase is in physical proximity to the promoter.

Answers:

A. Recall and short answer

1. Polycistronic operons group functionally-related genes such as all genes required for lactose catabolism or all genes needed for tryptophan synthesis. Grouping these genes that must all be transcribed and translated together to accomplish a task and placing them under a single regulatory system is more efficient than having separate, independently regulated genes.

2. The continued high levels of β–galactosidase activity in the face of rapidly declining *lac* mRNA levels reflect the relative stability of the β–galactosidase enzyme compared to the mRNA.

3. Regulation of the *lac* operon is tight but not perfect. Even without induction, *lac* mRNA levels are detectable at roughly 0.1% of fully induced levels. Also, there is another transporter that is constitutively expressed that carries some lactose into the cell.

4. The *lacO^c* allele is a mutant form of the *lac* operator that prevents or reduces repressor binding. This leads to constitutive expression of the *lac* operon.

5. *cis*-acting sites are elements of regulatory DNA sequences that must be physically linked to the genes they regulate. They do not produce a diffusible product but instead serve as binding sites for proteins that influence RNA polymerase's ability to initiate transcription. If the element is moved, the protein bound to the *cis*-acting site can no longer influence RNA polymerase's interaction with the promoter because the key spatial relationships between regulatory protein and RNA polymerase are altered.

6. The phenotypes of *lacI$^-$* and a *lacI$^-$* and *lacO^c* mutant are indistinguishable. Both mutations lead to constitutive expression of the *lac* operon. Although the phenotypes are the same, there are two genetic means to distinguish these mutations. Mapping the mutations will show if they colocalize or map different sites (the operator and the *lacI* repressor gene). Another way is to attempt complementation with the wild type allele. In the case of a *lacO^c* allele, complementation will not occur, as this allele is *cis*-dominant to the wild type *lacO$^+$* allele. In contrast, for a *lacI$^-$* allele, the wild type *lacI$^+$* allele will complement the defect. Here, the wild type allele acts in *trans* and is dominant to the *lacI$^-$* mutant allele. The different outcomes lie in the fact that *lacO* does not produce a diffusible product whereas *lacI* does.

7. *LacI* mutations that affect the DNA binding site (*LacI^{-d}* alleles) constitutively express the *lac* operon and act as dominant negative alleles. *lacI^s* alleles never express the *lac* operon (this is the nature of the designation: *lacI^s* derives from "super-repressor") and usually act as recessive alleles.

8. Induction of the *lac* operon in this partial diploid would occur inefficiently or not at all. This is because of the phenotype of all *lacI^s* mutants (no induction) and the dominant negative nature of the particular allele. For an allele to be dominant negative implies that it exerts its effects in the presence of the wild type allele.

9. A *lacO^c lacI^s* double mutant would constitutively express the *lac* operon. This is because the *lacO^c* allele will not bind repressor with a normal DNA binding site. Since the *lacI^s* mutations leaves the DNA binding site of the repressor intact, this mutant repressor would not bind the *lacO^c* form of the operator.

10. It is not surprising that the operator is palindromic because each dimer of the repressor contains two identical DNA binding sites oriented as mirror images of each other. One repressor dimer binds each operator, thus one expects that the operator will show dyad symmetry to accommodate the paired and oppositely orientated DNA binding sites of the repressor.

11. The very slow dissociation of repressor from the operator makes this model implausible. Induction would take far too long if repressor only bound inducer when inducer dissociated from the operator.

12. The region of repressor that encompasses the DNA binding domains undergoes the most radical change upon inducer binding. This region is moved significantly relative to the core so that the DNA binding domains no longer interact with the operator.

13. The repressor must bind *O1* and either *O2* or *O3*. Elimination of *O2* or *O3* has modest effects on repression because repressor will bind to *O1* and whichever other operator remains. Simultaneous deletion of *O2* and *O3* leaves only *O1* for repressor binding bound *O1* and this is not sufficient for effective repression.

14. This is possible because there are two stages of RNA polymerase's interaction with the promoter: closed and open complex formation. Binding RNA polymerase to the promoter (closed complex formation) is necessary but not sufficient for transcription initiation. Closed complex formation is enhanced when repressor is bound to the operators. Although RNA polymerase is bound to the promoter in a closed complex, repressor inhibits formation of the open complex. This is the stage at which repressor inhibits transcription initiation.

15. This would weaken regulation leading to "leaky" or constitutive transcription of the operon. The reason is that if repressor increased its affinity for random sequences, these would titrate repressor away from the operator.

B. Analytical and critical thinking

1. These studies provided early evidence about the function of *lac* operon components. The reason the *lacO⁺* allele did not complement the *lacOᶜ* allele is that that operator does not produce a diffusible product – it is a *cis*-acting element. In this case, the *lacOᶜ* allele is *cis*-dominant. The *lacI⁺* allele complements the *lacI⁻* allele because the *lacI* gene produces a diffusible product. Therefore, the *lacI* gene is *trans*-acting.

2. There are two reasons for this result. First, the *lacI⁻ᵈ* allele produces a repressor defective in DNA-binding. Since the repressor functions as a tetramer, mutant repressor protein will form multimers with the wild type protein produced from the introduced *lacI⁺* allele. These mixed multimers, like the mutimers formed only from the *lacI⁻ᵈ* product, are defective in DNA binding. Therefore, the mutant phenotype is not complemented and the partial diploids remain constitutive for *lac* operon expression. The *lacI⁻ᵈ* allele is said to be dominant negative because its associated phenotype is dominant to the wild type and it acts in a negative way with regard to repression.

3. In the presence of repressor/operator complex, roughly 99.7% of *lac* promoters would be occupied by an RNA polymerase. In the absence of repressor binding to the operator, roughly half of *lac* promoters would be occupied. This comes from rearrangement of the equation of

figure 10.22: [bound RNA polymerase]/[free RNA polymerase] = (K_A x [DNA]). There are 5.8 x 10^{-23} moles (i.e. 35 nucleotides/6.02 x 10^{23} nucleotides M^{-1}) of nucleotides in the promoter of each cell. Therefore, the concentration is 5.8 x 10^{-23} moles/1 x 10^{-15} liters = 5.83 x 10^{-8} M. In the case of the operator with bound repressor, [bound RNA polymerase]/[free RNA polymerase] = (2.5 x 109 M^{-1}) x (5.83 x 10^{-8} M) = 145. The fraction of promoter bound by polymerase is: [bound RNA polymerase]/[total RNA polymerase] (i.e., bound + free) = 145/146 = 0.993. For the case of operator not bound to repressor, [bound RNA polymerase]/[free RNA polymerase] = (1.7 x 107 M^{-1}) x (5.83 x 10^{-8} M) = 1.1. This indicates that approximately half of promoters would be bound by RNA polymerase.

4. If there were one repressor molecule and one operator in 3 x 10^9 bp of DNA, then the operator would be occupied 0.33% of the time. This is the ratio of the binding affinity for operator relative to random sequence DNA (1 x 10^7) relative to the number of non-specific binding sites (3 x 10^9), or (1 x 10^7)/(3 x 10^9) = 0.0033. This means that 1 – 0.0033, or 0.9967, of the time the operator is unoccupied. If there were 10 repressors in the cell, then the fraction of time the repressor would be <u>un</u>occupied would be $(0.9967)^{10}$ = 0.9674. This means that 1 – 0.9674, or 0.0326 of the time the repressor is occupied. Correspondingly, if there were 100 repressors per cell, then the fraction of time an operator is <u>un</u>occupied is $(0.9967)^{100}$ = 0.718. The fraction of time an operator is occupied is 1 – 0.718, or 0.282. Clearly as genome size increases, the number of repressors also must be increased.

C. Extensions and applications

1. If mouse genes could be put under tight regulation by the *lac* repressor/operator system, one could turn genes on or off at will. This would provide a powerful tool for fundamental studies of gene function during development and under different physiological conditions.

2. Three operators are present to simulate the situation found in *E. coli lac* operon.

3. This is a gamble as the repressor interacts in a complex way with *E. coli* RNA polymerase. It's not simply a case that of repressor excluding RNA polymerase from the promoter. Instead, repressor promotes RNA polymerase binding to form a closed complex and inhibits conversion from a closed to an open complex. Whether repressor could act similarly with the mammalian polymerase or act in a different way to block access to the promoter is not at all certain.

4. A *lacI* construct that expresses repressor in the mouse must also be introduced.

5. The W (wild type) form appears most effective as the degree of repression is greatest. (Degree of repression is the fold difference between the induced and repressed states.)

6. The "Tyrosinase transgenic" mouse is pigmented because it expresses tyrosinase and synthesizes melanin (pigment).

7. The "Tyr^{lacO}, lacI^R" mouse is white because repressor produced from expression of the lacI^R construct represses expression of the modified tyrosinase gene.

8. The "Tyr^{lacO}, lacI^R + IPTG" mouse is pigmented because the IPTG present in drinking water acted to remove the repressor from the operators of the engineered Tyr^{lacO} gene, thus allowing expression of tyrosinase and melanin synthesis.

D. Standardized test format

1. C
2. D
3. A
4. A
5. C
6. A
7. A
8. C
9. D
10. C
11. D
12. E
13. C
14. A
15. D

Chapter 11

Regulatory circuits

General Concepts	Section	Related Questions
Positive and negative transcriptional control of gene expression	11.1	
Regulation can occur at transcription initiation, transcription termination or translation	11.1	
Regulation is always mediated by a regulator interacting with a gene or mRNA	11.1	

Processes/structures	Section	Related Questions
Distinguishing positive and negative control	11.2	A1, A2, A3, B3, B4, C2, D8
Glucose repression	11.3	A4, D5, D9, D10
CRP and cAMP	11.4-11.6	B1, B2, C1, D5, D9, D10
Stringent control	11.7-11.9	A10, B2, D5
Autogenous regulation	11.10-11.13	A6, A7, B5
Attenuation	11.14-11.17	A9, B6, D1, D2, D8
Antisense RNA and small RNA's	11.18-11.21	A5, A8, D3, D4
RNA interference	11.22	D6

Questions:

A. Recall and short answer

1. What is the small molecule called that modulates the ability of a transcriptional activator to bind DNA?

2. What does the term "derepressed" mean?

3. What are the two ways in which induction of gene expression can be achieved?

4. Explain why the term "glucose repression" is a misnomer.

5. What are the four ways in which a small RNA can function in regulating gene expression?

6. What is the central mechanistic difference between autogenous control of ribosomal protein expression and autogenous control of tubulin expression?

7. What is the central mechanistic similarity between the way ribosomal proteins and tubulin exert autogenous control of their own expression.

8. Define the term antisense RNA.

9. What is attenuation?

10. Define the term "stringent response."

B. Analytical and critical thinking

1. The activation of sugar catabolism operons by CRP and cAMP depends on the level of cAMP, which is inversely correlated with the availability of glucose. PTS, a membrane localized sugar transport and phosphorylation system, phosphorylates a regulatory protein called IIA^{Glc} when glucose is unavailable. Phosphorylated IIA^{Glc} activates adenylate cyclase, increasing cAMP level, which in turn activates CRP, and induces the operons that are targets of CRP activation. What would the activation state be of CRP-regulated operons in a strain carrying a knockout mutation of IIA^{Glc} if glucose were supplied to the mutant?

2. Compare and contrast the effects of cAMP and pppGpp on gene expression.

3. Imagine that your favorite gene (*yfg1*) is expressed only when the bacterial strain you are studying is subcultured onto freshly prepared growth medium. Expression then decreases as the culture grows. A deletion of another gene, *yfgR*, abolished *yfg1* expression. Describe all the possible ways in which *yfgR* might function as a regulator of *yfg1* expression.

4. Suppose that your favorite gene (*yfg1*) is expressed when the bacterial strain you are studying is subcultured onto freshly prepared growth medium and expression decreases as the culture grows, just as in question B4. In this instance, deletion of *yfgR* causes *yfg1* to be expressed constitutively. Describe all the possible ways in which *yfgR* might function as a regulator of *yfg1* expression.

5. Does autogenous control always involve repression of gene expression by the product of the regulated gene? If not, give an example of the alternate type of regulation. (Hint: It may be necessary to review chapter 12 to answer this question.)

6. Compare and contrast the mechanisms of attenuation in the *E. coli* Trp operon and the *B. subtilis* Trp operon.

C. Extensions and applications

1. Describe the theory behind the analysis of DNA bending induced when a protein is bound to its operator.

2. If you had access to a DNA fragment containing the promoter, operator, and coding sequence for a gene that you knew was regulated at the level of transcriptional initiation, describe how you would determine if the operator is the binding site for a repressor or an activator.

D. Standardized test format

1. Which statement best describes the concept of "attenuation" as a method for regulating gene expression?

    A. Attenuators regulate transcriptional initiation.
    B. Attenuators regulate transcriptional termination.
    C. Attenuators regulate translational initiation.
    D. Attenuators regulate translational termination.
    E. none of the above

2. What is the function of anti-TRAP?

    A. It blocks expression of TRAP.
    B. It enhances TRAP expression.
    C. It degrades TRAP.
    D. It works opposite to TRAP by binding to the *trp* mRNA blocking formation of a transcriptional termination stem-loop.
    E. It binds to TRAP thereby preventing TRAP from causing transcriptional termination of the *trp* genes.

3. What is the function of the protein called "Dicer"?

    A. a protease that cleaves a preprotein into its functional form
    B. a nuclease that cleaves single-stranded RNA
    C. a nuclease that cleaves the precursor of a microRNA into its functional form
    D. a DNA binding protein that regulates gene expression
    E. a DNA binding protein involved in post-transcriptional gene silencing

4. Which of the following are activities of Dicer?

    A. RNA nuclease and helicase
    B. protease and RNA nuclease
    C. DNA nuclease and topoisomerase
    D. RNA nuclease and topoisomerase
    E. single stranded RNA binding and nuclease

5. What do cAMP and ppGpp have in common?

    A. Both are synthesized from GTP.
    B. Both are synthesized by idling ribosomes.
    C. Both are synthesized by adenylate cyclase.
    D. Both act as signals of amino acid starvation.
    E. Both act as signals of nutrient starvation.

6. Which of the following statements best describes the phenomenon of "cosuppression"?

    A. a transcriptional repressor that must be activated by a second protein called the cosuppressor
    B. expression of an antisense transcript causes more than one gene to be silenced
    C. expression of a transgene causes the expression of the endogenous copy of the gene to be silenced
    D. a transcriptional repressor that inhibits expression of a number of genes that are part of a regulon
    E. a small RNA that inhibits translation of more than one gene

7. Which of the following statements about the *E. coli trp* operon is false?

    A. Deletion of the *trp* leader increases both the basal and induced expression levels by about 10 fold.
    B. The *trp* leader is translated into a small enzyme that is needed for Trp synthesis.
    C. Expression of the *trp* genes is regulated by control of transcription initiation and termination.
    D. The *trp* operon produces a polycistronic mRNA with 5 cistrons.
    E. Regulation of transcriptional initiation is the primary mechanism for controlling *trp* operon expression; attenuation plays a lesser role.

8. Which of the following statements about ppGpp is false?

    A. ppGpp is degraded by the relA protein.
    B. ppGpp inhibits transcription initiation at rRNA promoters.
    C. ppGpp inhibits transcription elongation of rRNA transcripts.
    D. ppGpp is synthesized by RelA on idling ribosomes.
    E. ppGpp slows the growth rate of *E. coli*.

9. Which of the following statements about regulation mediated by CRP is false?

    A. The operator sequence for CRP binding can be found in the *gal*, *lac*, and *ara* operons.

    B. The operator for CRP binding is always located immediately 5' of the –35 sequence of the promoter.

    C. A complex of CRP and cAMP binds to the operator, CRP alone cannot bind to the operator.

    D. The form of CRP that is able to bind the operator is a dimer.

    E. The operator for CRP binding is a palindrome.

10. Which of the following most convincingly shows that CRP activates transcription by directly interacting with RNA polymerase?

    A. The CRP operator can activate transcription even if it is some distance from the promoter as long as it is positioned on the same face of the double helix as RNA polymerase.

    B. A mutation in the RNA polymerase alpha subunit is unable to be activated by CRP.

    C. CRP-dependent promoters are intrinsically poor promoters, meaning that RNA polymerase cannot efficiently initiate transcription without CRP.

    D. When CRP binds to its operator it bends the DNA 90°.

    E. A and B both indicate that CRP interacts with RNA polymerase.

Answers:

A. Recall and short answer

1. inducer

2. Derepressed has the same meaning as induced, but refers to induction of a repressible system.

3. Induction can be achieved by inactivation of a repressor or activation of an activator.

4. When bacteria are presented with glucose and another carbon source, glucose will be preferentially catabolized. The operons encoding enzymes for catabolism of the alternate carbon source are usually not repressed by glucose, rather they are activated when glucose is unavailable.

5. By base pairing with the target RNA the regulator can prevent the binding of proteins that are able only to bind single stranded RNA or the regulator can prevent binding of ribosomes thereby inhibiting translation, or the regulator can prevent the formation of an alternative base pairing conformation or the regulator can base pair with the target RNA, forming a dsRNA that is targeted for degradation by endonucleases.

6. When ribosomal proteins inhibit their own expression they bind to their mRNA preventing the initiation of translation, whereas when tubulin inhibits its own expression it initiates the degradation of its own mRNA.

7. In both instances the free forms of the proteins are active in the regulation system. Free ribosomal proteins accumulate when they are synthesized in excess over other ribosomal components. Tubulin monomers accumulate when they are synthesized faster than the monomers can be assembled into microtubules.

8. Antisense RNA is an RNA molecule that is transcribed using the sense strand of DNA as the template. The molecule is the RNA equivalent of the template DNA strand.

9. Attenuation is the process of transcriptional termination that serves as a control mechanism to regulate the expression of genes located downstream of the attenuation site.

10. The "stringent response" refers to the ability of bacteria to down-regulate a variety of growth processes when they encounter poor growth conditions.

B. Analytical and critical thinking

1. The operons would be in the repressed state because cAMP levels would be low. However, the IIA^{Glc} mutation is irrelevant to the repressed state of the operons under the growth conditions because IIA^{Glc} would only be functional if glucose were absent.

2. cAMP and pppGpp are derived from nucleotides and can be considered signaling molecules that indicate the poor growth conditions, cAMP signals the lack of glucose and pppGpp signals the lack of a sufficient supply of amino acids. The enzymes that synthesize these compounds are activated by the nutrient limiting condition. cAMP is an inducer of the transcriptional activator CRP. By contrast pppGpp represses the expression of genes by interacting directly with RNA polymerase.

3. Since *yfg1* is not expressed when *yfgR* is absent it can be concluded that the *yfgR* product is necessary for *yfg1* expression. There are two ways in which *yfgR* could be required for *yfg1* expression. First, the *yfgR* protein could be an activator in a positive control system. Normally, the activator stimulates *yfg1* expression immediately after subculture, but without it *yfg1* gene cannot be expressed. Alternatively, the *yfgR* protein or a molecule synthesized by the *yfgR* protein could function as the inducer of an activator in a positive control system or the inducer of a repressor in a negative control system. Normally, the inducer either activates an activator after subculture or inactivates a repressor, thereby stimulating *yfg1* expression, but without the inducer *yfg1* cannot be induced or remains repressed after subculture.

4. Since *yfg1* is expressed constitutively when *yfgR* is absent it can be concluded that the *yfgR* product is required to block *yfg1* expression. There are two ways in which *yfgR* could repress *yfg1* expression as either the repressor itself or as a corepressor that regulates the repressor's activity. First, the *yfgR* protein could be a repressor in a negative control system. Normally, the repressor blocks *yfg1* expression, but in its absence *yfg1* expression remains high even in the later stages of the culture cycle. Alternatively, the *yfgR* protein or a molecule synthesized by the *yfgR* protein could function as the corepressor in a negative control system or as the corepressor in a positive control system. Normally, the corepressor is necessary for a repressor to function in blocking *yfg1* expression or to inactivate an activator. But in the absence of the corepressor the repressor is unable to repress *yfg1* expression or is unable to inactivate the activator. In both cases the *yfg1* gene cannot be repressed or is constitutively induced.

5. No, autogenous control can also function to stimulate expression of a gene. A prominent example is the way in which lambda cI protein positively regulates its own transcription when *cI* is bound to O$_{R2}$. In this instance, cI functions as a positive regulator of gene expression at the P$_{RM}$ promoter.

6. In the *E. coli* Trp operon the attenuator function is determined by the position of ribosomes translating the leader of the mRNA. When Trp is available ribosomes translate all the way through the leader, reaching the termination codon. In this configuration the transcriptional termination stem-loop can form. Alternatively, when Trp is limiting, ribosomes stall at Trp codons in the middle of the leader. In this configuration the transcriptional termination stem-loop fails to form and RNA

polymerase transcribes the *Trp* genes downstream of the terminator. In the *B. subtilis* the attenuator function is determined by whether a protein called TRAP binds to the mRNA. When Trp is available TRAP binds to the mRNA and its binding promotes the formation of the transcriptional termination stem loop. When Trp is limiting, TRAP doesn't bind and the transcriptional termination stem loop fails to form resulting in expression of genes downstream of the terminator. Therefore, both attenuators depend on *trans*-acting factors that promote or inhibit the formation of a transcriptional termination stem loop. In the *E. coli* system the *trans*-factor is a translating ribosome, in the *B. subtilis* system the *trans*-factor is an RNA binding protein.

C. Extensions and applications

1. When a bend exists in DNA its mobility in an electrophoretic gel is reduced. DNA bending by a DNA binding protein is analyzed by assaying DNA fragments of the same size in which the operator is located at varying positions. When the protein introduces a bend in the center of the fragment, its mobility will be slower than if the bend exists at the ends of the fragment. If the protein does not introduce a bend, the DNA fragments will have the same mobility regardless of where the protein is bound.

2. There are several possible answers. However, the most direct is to delete the operator sequence and introduce the modified gene into cells. If expression of the gene is constitutive, the operator was for a repressor. If the gene is not expressed, it could mean that the operator was for an activator, or it could mean that the deletion has destroyed the promoter. A better experiment would be to place the operator near another gene. If the expression of this reporter gene is activated, one could conclude that the operator is the binding site of an activator.

D. Standardized test format

    1. B
    2. E
    3. C
    4. A
    5. E
    6. C
    7. B
    8. A
    9. B
    10. E

Chapter 12

Phage strategies

General Concepts	Section	Related Questions
Phage lifestyles	12.1	
Lytic development	12.2-12.5	
Lysogenic development	12.6	

Processes/structures

	Section	Related Questions
Cascades of gene expression control lytic growth	12.3-12.4	A8, B4, D6
Two promoters control the early and delayed early genes of lambda	12.8-12.15	B1, C1
The early and delayed early genes are needed for both lytic growth and lysogeny	12.6	C1, C2, D2, D3, D4
cI repressor is essential for lysogeny	12.8-12.15	A7, A10, B1, B3, C1, C2, D7, D8, D9
cro repressor is needed for lytic growth	12.19	C1, D9
N antiterminator is needed for both lytic growth and lysogeny	12.7	C1, D1, D5
Q antiterminator is needed for late gene expression	12.7	D6
cII and *cIII* work together to establish lysogeny	12.16-12.17	A9, B1, C2

Questions:

A. Recall and short answer

1. Define the term "prophage."

2. A phage that always lyses its host is described as being _____.

3. How does the life cycle of a lysogenic phage differ from a lytic phage?

4. List 3 strategies used by phage to control the expression of delayed early genes.

5. Compare and contrast the ways in which lytic and lysogenic phage genomes are replicated.

6. Compare and contrast the way in which phage genomes and episomes are transferred from one host to another.

7. *cI* protein can function as either a negative or a positive regulator of transcription. Describe how.

8. Lytic development is controlled by a _____ of gene expression. The _____ genes are expressed first, followed by the _____ and finally the _____.

9. Describe how *cIII* controls gene expression without directly contacting DNA.

10. How would mutation of O_R2 affect the outcome of lambda infection and the stability of a lysogen derived from the mutant phage?

B. Analytical and critical thinking

1. Explain the phenotypes of *cI*, *cII* and *cIII* mutants. *cI* mutants never form lysogens. By contrast, although the frequency of lysogen formation is reduced in *cII* and *cIII* mutants, these mutants do rarely form lysogens that are as stable as wild type lysogens.

2. Give a plausible reason why scientists used the term "lysogen" to describe a bacterial cell that contains a prophage. Since the prophage is quiescent this term doesn't quite make sense, does it?

3. If a lysogenic culture always contains a low titer of phage why aren't the bacterial cells themselves lysed by the infective particles?

4. Speculate on the functional reason why genome replication proteins are expressed during the early period after infection, while capsid and assembly proteins are expressed late?

5. List some functions that distinguish lytic phage like T4 and T7 from lysogenic phage such as lambda.

C. Extensions and applications

1. Imagine that a lambda phage was constructed that carried a conditional mutation in *cI*, and knockout mutations in *N* and *O*. Such a phage would not be able to grow lytically because of the mutations in *N* and *O*. Yet it would be able to form a lysogen (although rarely due to the *N* deletion) if infection was carried out at the permissive condition. How would a lysogen of this specially constructed mutant respond to superinfection with a wild type lambda phage? How would the lysogen respond to superinfection if it were exposed to the non-permissive condition and then returned to the permissive condition before being infected with the wild type phage?

2. About 1000 fold more *cI* repressor is expressed in cells newly infected by lambda compared with the level in a lysogen. If the function of *cI* is to block lytic growth why is so much more repressor needed soon after infection than after the lysogen is formed?

D. Standardized test format

1. What is the function of the *N* protein?

    A. transcription termination
    B. transcription antitermination
    C. translation termination
    D. translational antitermination
    E. replication termination

2. What is a plaque assay?

    A. the growth of phage observed as small colonies growing on a petri dish
    B. the growth of phage observed as clear zones of killed bacterial cells
    C. a way to measure the number of phage by viewing them with an electron microscope
    D. a way to measure the frequency of lysogen formation
    E. a way to measure the number of host cells killed by an infection

3. How does the plaque formed by a lytic phage differ from that formed by a lysogenic phage?

    A. The lytic phage forms clear plaques and the lysogenic phage forms turbid plaques.
    B. The lysogenic phage forms clear plaques and the lytic phage forms turbid plaques.
    C. There is no difference between the plaques formed by lytic and lysogenic phage.
    D. Lytic phage form plaques but lysogenic phage do not.
    E. Lytic phage form plaques with ragged edges.

Mini-Matching (match the lambda phage gene or locus (4 to 8) with the description of its null mutant phenotype, one response each)
4. *O*
5. *N*
6. *Q*
7. O_R and O_L
8. *CRO*

    A. This mutant phage cannot synthesize head and tail proteins so it cannot grow lytically, but it can form a lysogen.
    B. This mutant phage is unable to replicate as an extrachromosomal element so it cannot grow lytically, but it can form a lysogen.
    C. This forms lysogens more frequently than wild type.
    D. This mutant phage has a high frequency of failed infection, where neither lytic nor lysogenic growth occurs.
    E. This mutant phage can lyse a lambda lysogen.

9. What is the function of the helix-turn-helix domain of *cI*?

    A. formation of the dimer form of the *cI* protein
    B. interaction with RNA polymerase
    C. operator binding
    D. lytic growth
    E. transcriptional antitermination

10. *CRO* and *cI* are homologous genes, meaning that they show sequence conservation and evolved from a common ancestor. In the absence of their nucleotide sequences what evidence best indicates that the two genes are evolutionarily related?

      A. The two genes are next to each other on the lambda chromosome.
      B. *CRO* and *cI* proteins bind to the same operator sites.
      C. *CRO* and *cI* are both lambda genes.
      D. Both proteins are transcriptional repressors.
      E. Both proteins are about the same molecular weight.

Answers:

A. Recall and short answer

1. a phage genome that exists in a quiescent state within a bacterium

2. lytic or virulent

3. Infection with a lytic phage always results in lysis of the host, whereas infection with a lysogenic phage can result either in lysis of the host or the formation of a lysogenic host.

4. One strategy is that an early gene encodes an RNA polymerase that specifically recognizes the promoter for the delayed early genes. Another strategy is that an early gene encodes a sigma factor that modifies the activity of the hosts RNA polymerase, directing it to specifically recognize the promoter of the delayed early genes. A third strategy is that an early gene encodes an antitermination factor that allows the host RNA polymerase to transcribe past a Rho-dependent terminator located ahead of the delayed early genes leading to their expression.

5. The lytic phage genome is always replicated independently of the host bacterial chromosome. Replication occurs via the bidirectional method and the rolling circle method. The lysogenic phage genome can replicate either as an extrachromosomal DNA element or as a locus of the host chromosome. As an extrachromosomal element it replicates via the bidirectional and the rolling circle methods, and as a locus of the bacterial chromosome it replicates via the bidirectional method.

6. Phage genomes are packaged into a protein shell called a capsid. The capsid functions as a structure for delivery of the genome into a host upon infection. Episomes are transferred by conjugation, a process by which the genome is directly transferred from one host cell to another.

7. An example of negative control is when *cI* is bound to O_R1, it blocks the ability of RNA polymerase to initiate transcription from the P_R promoter. An example of positive control is when *cI* is bound to O_R2, it enhances the ability of RNA polymerase to initiate transcription at the P_{RM} promoter.

8. cascade, early, delayed early, late

9. *cIII* protects *cII* from degradation by host proteases. *cII* contacts DNA and controls gene expression.

10. O_R2 is responsible for autogenous, positive control of *cI* expression from the lysogen maintenance promoter P_{RM}. So mutation of O_R2 would not effect the ability to form a lysogen, however, since positive control is necessary to produce sufficient *cI* to maintain the lysogen, the mutant prophage would be more easily induced than a wild type.

B. Analytical and critical thinking

1. *cI* is required for both lysogen formation and lysogen maintenance. If *cI* is not available to repress transcription from P_R and P_L, lysogens will not form. *cII* and *cIII* are required only for the formation of lysogens, but not for maintaining them.

2. At a very low frequency, individual prophage in the lysogenic culture become induced resulting in the release of infective phage particles into the growth medium. Therefore, a lysogenic culture always produces a low titer of infective phage particles that can be detected by infecting an appropriate bacterial host, hence, the term "lysogenic," *lysis-producing*.

3. The lysogenic cells are immune to infection by phage of the same type. Since the infective phage in the lysogenic culture are derived from the prophage harbored within the lysogenic cells, they are genetically identical to the prophage.

4. To maximize the yield of phage. Phage genomes serve two functions, as templates for replication and as substrates for packaging into capsids. During the earliest phase of growth the number of phage genomes is amplified. Later, after many genome copies have accumulated, the capsid and assembly proteins have excess genomes to package. If capsid and assembly proteins were produced too early after infection, the packaging would decrease the number of genomes available for replication reducing the overall yield of infective phage particles produced.

5. A successful outcome of lytic phage infection is to maximize the yield of progeny, whereas lysogeny is a possible outcome of infection with a lysogenic phage. Therefore, lytic phage encode functions intended to subjugate the host. For example, lytic phage express enzymes that interfere with host gene expression and break down the hosts DNA. Such functions are incompatible with lysogenic growth and are not present in a lysogenic phage. By contrast, lysogenic phage like lambda have more complex regulatory circuits that are needed for production and maintenance of the lysogenic state, this regulation doesn't exist in lytic phage.

C. Extensions and applications

1. Under the permissive condition *cI* protein would be active and would block lytic growth of superinfecting phage. So the lysogen would be immune. However, if the lysogen were raised to the non-permissive condition, *cI* repressor would become inactive, relieving repression of P_L and P_R resulting in expression of *CRO* and *N*. Since the *N* gene is deleted transcription would not occur past the terminators. The deleted *O* gene would ensure that any chance transcription through the terminators would not lead to lytic growth. The expressed *CRO* protein would bind to O_R3 thereby preventing *cI* expression from P_{RM} so *cI* expression would not be regained after return of the lysogen to the permissive condition. Thereafter, the lysogen will have lost immunity and will be sensitive to superinfection with wild-type phage.

2. Upon infection, lambda progresses toward both lytic and lysogenic growth at once. One of the initial events leading toward lytic growth is the replication and rapid accumulation of phage genomes. If lysogeny is to be the outcome *cI* repressors must gain control of each of these many genomic copies. By contrast, in the lysogen, there is only a single copy of the phage genome, so comparatively little repressor is needed to keep the prophage in check.

D. Standardized test format

1. B
2. B
3. A
4. B
5. D
6. A
7. E
8. C
9. C
10. B

Chapter 13

The replicon

General Concepts	Section	Related Questions
Replicons can be circular or linear	13.1-13.2	
The replicon is controlled by a cis-acting element called an origin	13.3-13.7	
DNA can be transferred from one bacterial cell to another by a process involving replication	13.12	
Regulation of replication is linked to the cell cycle	13.14	
Cell division is controlled	13.15-13.20	
Replicons must be partitioned	13.20-13.21	
Some replicons are incompatible	13.22-13.23	
Mitochondrial genomes are replicated	13.24	

Processes/structures	Section	Related Questions
Linear and circular replicons	13.2, 13. 4, 13.5	A1, A8, B1, D2, D3
Mapping origins	13.3	C1
Isolating origins	13.6	A10, B4, D4
Mitochondrial origins	13.7	A2
Linear viral origins	13.8	A9
Rolling circle replication	13.9-13.11	A3, A8
F-plasmid conjugation	13.12-13.3	A3, A7, B1, B2, D1, D5, D6, D11
Cell division	13.14-13.18	A5, A6, D6
Partitioning of replicons	13.19-13.21	A4, A3, D2, D7, D9, D10
Plasmid incompatibility	13.22-13.23	A4, D2
Mitochondrial replication	13.24	C2

Questions:

A. Recall and short answer

1. Define the term "replicon."

2. Define the term "D-loop."

3. Define the term "relaxase."

4. Define the term "copy number."

5. Define the term "Z-ring."

6. FtsZ is a _____ required for _____ formation in *E. coli*.

7. Replication at _____ initiates _____ replication during F-factor transfer.

8. Phage genomes can replicate through two different mechanisms: _____ and _____ .

9. The covalent linkage between the terminal protein and the the adenovirus genome is between a _____ residue of the terminal protein and the _____ end of the genome.

10. The sequence of the *Saccharomyces cerevisiae* origin of replication is _____ and is bound by a complex of 6 proteins called the _____ .

B. Analytical and critical thinking

1. Define the term "episome" and provide two examples of episome systems.

2. The map of *E. coli* genes is depicted as a 100-minute clock. Explain why.

3. Describe how RNA regulates replication of ColEI plasmids and plasmid compatibility.

4. What is the key similarity between D-loop type replication of mitochondrial genomes and rolling circle replication?

5. What is the problem of replication at the ends of genomes and how does the adenovirus overcome the problem?

C. Extensions and applications

1. Describe two techniques used to map origins of replication.

2. Assume that you are investigating the possibility of creating transgenic organisms by mitochondrial transformation. You have successfully demonstrated that a foreign gene can be integrated into the mitochondrial genome of cells from your experimental higher eukaryotic organism. You now are seeking to produce a whole organism from your transformed cell. What problems might you encounter in creating the whole mitochondrially transgenic organism?

D. Standardized test format

1. Which of the following statements about F-factor mediated DNA transfer is false.

    A. DNA is transferred from one cell to another in the single stranded form.
    B. DNA transfer occurs at the same time as replication via the rolling circle method.
    C. DNA transfer occurs through a structure called the F-pilus.
    D. Transfer occurs when TraY, a relaxase, cleaves one DNA strand at the origin.
    E. Host genes are transferred when F-plasmid is integrated into the host chromosome.

2. What is *oriC*?

    A. the replication origin of the *E. coli* chromosome
    B. the replication terminator of the *E. coli* chromosome
    C. a *trans*-factor required for initiation of replication at the chromosome origin.
    D. the gene encoding a factor required for termination of replication.
    E. a transcriptional regulator of genes required during oxidative stress.

Match the terms in #3 to #7 with the definition A to E.
3. Tus
4. ARS
5. *Hfr*
6. MinB
7. ParB

    A. an element derived from an origin of replication that confers the ability to replicate efficiently in yeast
    B. protein required for partitioning of single copy plasmids during cell division in *E. coli*
    C. *E. coli* strain carrying an integrated F-factor
    D. a protein that helps segregate plasmid copies to opposite daughter cells
    E. a protein required for placement of the septum during cell division in *E. coli*

Match the gene (#8 to #12) with the phenotype resulting from inactivation of the gene (A to E).
8. *ihf*
9. *parA*
10. *mukB*
11. *traJ*
12. *terD*

    A. Plasmid encoded gene, when mutated causes a failure to segregate properly.
    B. Mutation inactivates conjugation of an Hfr strain.
    C. *E. coli* mutants in which the chromosome is easily lost due to a failure to partition into daughter cells.
    D. There is no phenotype associated with this mutation.
    E. Host mutant in which phage lambda DNA cannot integrate.

Answers:

A. Recall and short answer

1. A replicon is the unit of DNA that is replicated and includes an origin of replication.

2. the region within mitochondrial DNA in which an RNA displaces one of the DNA strands and base pairs with the other

3. an enzyme that cuts one strand of DNA and binds to the free 5'-end

4. Copy number refers to the number of plasmids that exist in a cell. Each plasmid can act as an independent replicon.

5. the ring around the center of the *E. coli* cell where septum formation occurs and where the FtsZ protein localizes

6. GTPase, septum

7. oriT, rolling circle

8. bidirectional replication, rolling circle replication

9. serine, 5'-end

10. A-T-rich, ORC

B. Analytical and critical thinking

1. An episome is a DNA element that can exist as a free circular plasmid or may become integrated into the host chromosome. Two examples of episomes include the F-factor and certain phage DNA.

2. F-plasmids can integrate into many locations in the host chromosome. During conjugation, the time required for host genes to enter the recipient cell depends on their position relative to the F-factor integration site, the further the genes are from the site the longer it will take to enter the recipient strain. One particular Hfr strain was used for mapping. The map refers to the necessary transfer times from a particular Hfr strain for each gene and is determined by the position of the F factor within the genome.

3. Replication of ColEI replicons is initiated by formation of an RNA transcript near the origin. RNAse H cleaves the transcript while it remains hybridized to the origin generating the primer that is used to initiate replication. An anti-primer RNA called RNAI inhibits priming by base pairing with the primer RNA. RNAI also prevents replication of other ColEI plasmids, also by base pairing with the primer RNA.

4. In both cases replication is initiated and proceeds on only one strand.

5. Since replication always requires a primer and proceeds from 5' to 3', special mechanisms are required to replicate an entire linear genome. In essence, a mechanism is required to prime replication at the very end of the template strand, otherwise the genome will become progressively shorter after each round of replication. Adenovirus solves the problem by producing a terminal protein that binds to and becomes covalently linked to the 5' end of one strand. The terminal protein has a cytidine nucleotide linked to a specific serine residue that serves as the primer. DNA polymerase extends from the free 3' end of the cytidine nucleotide, copying the template strand linear genomes.

C. Extensions and applications

1. The first method relies on the ability to visualize replication origins as "eyes" on autoradiographic film. Replicating DNA units are labeled with radioactive deoxynucleotides and then are spread onto a surface that can be exposed to film. The second method relies on the aberrant migration of replication "eyes" in an electrophoresis gel.

2. Eukaryotic cells contain many mitochondria and it is very likely that only some of these mitochondria were the targets of transformation. The problem that must be overcome stems from the fact that mitochondrial replication and segregation is stochastic. The challenge will be to remove all non-transformed mitochondria from the transformed cell. Otherwise, the transgenic organism derived from the cell will be chimeric, and in fact, it is likely that some cells of the organism may not contain any transformed mitochondria.

D. Standardized test format

    1. D
    2. A
    3. D
    4. A
    5. C
    6. E
    7. B
    8. E
    9. C
    10. A
    11. B
    12. D

Chapter 14

DNA replication

General Concepts	Section	Related Questions
Polymerases synthesize DNA	14.1-14.5	B1, B2, D11, D12
Replication is semidiscontinuous	14.6	A2, A13, D6, B6
Replication requires a primer	14.6-14.8	A4, A5, D5, D13
Replication of both strands is coordinated	14.9-14.12	B4, D6
Eukaryotic and *E. coli* polymerases are similar	14.13	A8, A9, D11
Replication begins at replication origins	14.15-14.17	A6, D4, D14
Replication initiation is controlled	14.18-14.21	A10, A11, B3, C2, D9, D15
Processes/structures		
Replication of DNA	14.1-14.2	A1, A15, B2, D4
Removal of RNA primers	14.3	A1, A3, A14
Proofreading of replicated DNA	14.4	A1, D1, D3
Helicase function	14.7, 14.17	A2, D7, D8
Single-strand binding protein function	14.7	A2, D10
Generating primers	14.7, 14.8	A2, D5, D13
Coordinating activities at the replication fork	14.9, 14.11	B4, B6, C1, D6
Ligation	14.12	A2, A3, D2
Initiation of replication	14.15-14.21	A7, A11, B3, C2, D4, D15
Examples, Applications and Techniques		
E. coli DNA polymerases	14.1-14.6	A1, A2, A14, B5, C1, D12
φX174 replication initiation	14.7	A4, B6, D5
E. coli replication origin	14.8, 14.18	A6, A7, B3, D14
T4 replication	14.14	A12, D11
Lambda replication initiation	14.16	A7
Eukaryotic replication origins	14.20, 14.21	A7, A11, C2, D9

Questions:

A. Recall and short answer

1. What are the three enzymatic activities of DNA polymerase I of *E. coli* and what is the purpose of each activity?

2. What is the order of action of the *E. coli* proteins needed to replicate the lagging strand?

3. Identify two genes required for *E. coli* replication in which conditional lethal mutations are expected to lead to the accumulation of large numbers of Okazaki fragments.

4. The primer for cellular chromosome replication is a short RNA. Under what conditions does a nick in DNA serve as a primer?

5. In the replication of some linear viruses a priming protein is used. How do these proteins function as primers?

6. The protein responsible for the initial replication event at the *E. coli* replication origin is _____.

7. Two sequence features that are shared by many replication origins are _____ and _____.

8. What are two major differences between the replication and the repair polymerases in eukaryotic cells?

9. The two eukaryotic DNA polymerases required for leading strand synthesis are _____ and _____.

10. The proteins that control and prevent extra rounds of initiation of replication from eukaryotic replication origins are the _____.

11. What are the three proteins or protein complexes that are required for initiation of replication in yeast cells?

12. Why is it necessary for phage T4 to encode its own replication enzymes when phage φX174 and many others can make use of the host enzymes for replication?

13. What is an Okazaki fragment?

14. _____ is the process by which DNA polymerase I begins synthesis at the 3 OH group in a single-strand break by removing nucleotides from the 3 side of the break and by adding others to replace them onto the 5 side of the break. The activities required for this are its _____ and its _____ activities.

15. Would a conditional lethal *dna*B mutation that prevented helicase from unwinding the template DNA have a "quick-stop" or a "slow-stop" phenotype? What about a similar mutation in *dna*A or *pol*C?

B. Analytical and critical thinking

1. DNA polymerases have a common structure. What is the relationship of the structural domains to the function of the protein? What does the conserved structure imply about the evolution of DNA polymerases?

2. A typical polymerase chain reaction (PCR) includes a small amount of genomic DNA, the four deoxynucleotide triphosphates, and two short oligonucleotides. Why are each of these components needed in the reaction?

3. *E. coli* DNA is methylated on the N^6 position of the adenines in both strands of all GATC palindromes within the DNA duplex. The 245 bp *ori*C region has 13 GATC sites within it, is this fewer than, more than, or about the same as the expected number of sites predicted for this length of sequence if the DNA is a mixture of equal numbers of AT and GC base pairs? What happens to the methylation patterns of these sites as the DNA is replicated? How do changes of the methylation pattern after replication of the *ori*C region affect its ability to serve as an origin of replication?

4. Discuss the functions of the major subcomplexes in *E. coli* DNA polymerase III and their organization in the replisome.

5. Comparison of the nucleotide sequences of a wild type allele of a gene and of an allele of the same gene in an auxotrophic mutant revealed that one location within the coding region the wild type gene had 7 AT base pairs in a row but the mutant had only 6 AT base pairs. Propose a mechanism for generating the mutation based on a replication error and describe the consequences of the mutation on the mRNA and the peptide made from the mutant gene.

6. DNA replication in *E. coli* is described as semidiscontinuous; what does this mean? The φX174 virus has a circular genome and uses rolling circle replication to generate single strand genomes for the progeny viruses. Is φX174 rolling circle replication semidiscontinuous?

C. Extensions and applications

1. *E. coli* DNA polymerases vary in their ability to replicate DNA using damaged templates. They vary in their ability to use specific types of damaged templates as well as in the products synthesized in response to the damaged DNA. Goodman has recently reviewed the current knowledge on polymerases involved with error-prone repair (2002 *Annu. Rev. Biochem.* 71:17-50). In *E. coli* polymerases PolII, PolIV and PolV are induced by ultraviolet light exposure and have error-prone properties. PolII like PolIII and PolI is normally a high fidelity polymerase but it is known to generate -2 frameshifts in response to some guanine adducts and to copy abasic lesions if PolV is absent. PolII is induced very early after UV exposure but PolV induction does not occur until about 45 minutes after UV exposure. Replication forks are blocked by damage induced by UV. PolII is capable of error-free replication restart on the damaged templates, while PolV translesion synthesis is error-prone. PolV is capable of translesion synthesis of templates that contain thymine dimers, TT (6-4) photoproducts, or abasic sites. PolV typically adds a G opposite the 3′ T of the template TT (6-4) photoproduct. PolIV efficiently uses misaligned primer-template pairs to generate frameshift mutations. It is almost exclusively responsible for adaptive mutations where bacterial cells are placed in nonlethal but selective environments. All five DNA polymerases in *E. coli* bind to the same processivity clamp. Which of the other polymerases displaces PolIII at damaged sites depends on the timing, polymerase abundance, and specific DNA damage.

What advantage do the error-prone polymerases provide to cells? How does *E. coli* balance the advantages and disadvantages associated with the error-prone polymerases? Why is it advantageous to have PolII activity increased before PolV activity after UV exposure? What is the physical basis for the different specificities of the enzymes in terms of the types of damaged template that they are able to use and the types of replication products they are able to synthesize? Why is it reasonable for all the polymerases to interact with the same processivity clamp?

2. In eukaryotic cells the initiation of replication is tightly controlled so that it occurs in S phase and so that replication is limited to one round. The key to this control is the availability of certain proteins, licensing factors. These proteins become functional at the end of G1. Nishitani and coworkers (2001 *J. Biol. Chem.* 276:44905-44911) have studied the accumulation and turnover of human licensing factor hCtd1 during the mitotic cell cycle. HCtd1 is present in G1 and decreases rapidly in S phase. It is barely detectable early in S phase. Changes in the abundance of hCtd1 appear to be controlled by proteolytic rather than by transcriptional processes. The regulation of hCtd1 is different from that of a second human licensing factor HCdc6/18 that is present in similar abundance throughout the cell cycle except for a brief decrease at the end of mitosis. HCtd1 like HCdc 6/18 is required for loading the MCM complex on to chromatin. Proteins homologous to hCtd1 are present in some but apparently not all eukaryotes. During the transition from G1 to S the protein kinases CDK-cyclin and Cdc7 activate the licensed origins and Cdc45 and DNA polymerase bind to initiate replication.

The role of the origin recognition complex has been best studied in *Saccharomyces cerevisiae*; what is its function? Are similar proteins present in human cells? How are the availability and function of some licensing factors tied to nuclear division during mitosis? Phosphorylation is used to control the activity of many of the proteins involved in cell cycle control and in replication initiation; what happens when a protein is phosphorylated? Why does it make sense for a cell to regulate two essential licensing factors by different processes? What process ensures that reinitiation of replication does not take place in human cells?

D. Standardized test format

1. One consequence of mutating the $3' - 5'$ exonuclease of *E. coli* DNA polymerase I is to

   A. decrease the mutation rate of the cell.
   B. increase the mutation rate of the cell.
   C. prevent the removal of the RNA primers during lagging strand replication.
   D. decrease the processivity of the enzyme.
   E. decrease the ability of the enzyme to displace DNA polymerase III during lagging strand replication.

2. DNA ligases catalyze phosphodiester bond formation via a two step process.

   A. In the first step an AMP group is added to the nucleotide on the $3'$ side of the nick and the second step the phosphodiester bond is made and the AMP group released.
   B. In the first step an AMP group is added to the nucleotide on the $5'$ side of the nick and in the second step the phosphodiester bond is made and the AMP group released.
   C. In the first step a GMP group is added to the nucleotide on the $5'$ side of the nick and in the second step the phosphodiester bond is made and the GMP group released.
   D. In the first step a GMP group is added to the nucleotide on the $3'$ side of the nick and in the second step the phosphodiester bond is made and the GMP group released.
   E. In the first step a CMP group is added to the nucleotide on the $3'$ side of the nick and in the second step the phosphodiester bond is made and the CMP group released.

3. The proofreading function of DNA polymerases III is its

    A. 5′ – 3′ polymerase activity.
    B. 5′ – 3′ exonuclease activity.
    C. primase activity.
    D. ligase activity.
    E. 3′ – 5′ exonuclease activity.

4. The *E. coli* chromosome replicates bidirectionally, this means that there

    A. are two replication origins on the bacterial chromosome.
    B. are two daughter duplexes formed after replication of the DNA.
    C. are two replication forks established that move in opposite directions.
    D. is replication of both a leading strand and a lagging strand.
    E. is synthesis of DNA by both DNA polymerase I and DNA polymerase III.

5. DNA polymerases all require a 3′ OH group to add nucleotides to a replicating strand. In rolling circle replication of φX174 how is the 3′ OH group generated?

    A. A viral RNA polymerase makes a short primer that carries the 3′ OH group.
    B. The cellular primase makes a short primer that carries the 3′ OH group.
    C. A priming protein brings in a nucleotide that carries the 3′ OH group.
    D. The viral A protein makes a nick in one DNA strand to form the 3′ OH group.
    E. A helicase makes a nick in one DNA strand to form the 3′ OH group.

6. Lagging strand replication in *E. coli* is discontinuous because

    A. the antiparallel nature of the template DNA strands allows only one strand to be replicated continuously, lagging strand replication in the opposite direction to the unwinding of the parental duplex can occur only as template becomes available.
    B. it takes place after the replication of the leading strand has occurred and uses a different polymerase subunit in the replisome.
    C. the processivity of DNA polymerase III is reduced when it is replicating the lagging strand.
    D. RNA primers must be synthesized by the primosome at many locations during replication of the lagging strand.
    E. the processivity of DNA polymerase I is low, therefore replication must begin at many origins in order to replicate the entire genome within the time available.

7. Which protein separates the strands of the parental duplex DNA in order to prepare single-stranded template DNA for replication?

    A. a topoisomerase
    B. the single-strand binding protein
    C. primase
    D. helicase
    E. DNA polymerase III

8. How does the Tus protein stop the movement of a replication fork in *E. coli*?

    A. The Tus protein has a contra-helicase activity that stops the DnaB protein from unwinding the template duplex.
    B. The Tus protein has an anti-primase activity that prevents synthesis of new primers.
    C. The Tus protein has an exonuclease activity that removes nucleotides from the 3´ end of the replicating DNA strand.
    D. The Tus protein binds to DNA polymerase III and inhibits the ability of the polymerase to add nucleotides to the replicating DNA strand.
    E. The Tus protein generates strand breaks in the single-stranded template DNA that stall further replication.

9. The function of the eukaryotic origin recognition complex (ORC) is to

    A. regulate replication to ensure that replication begins simultaneously at all of the origins in the genome of the cell.
    B. bind to the replication origins and in turn facilitate the binding and action of the Cdc6 protein and licensing factors that control eukaryotic replication.
    C. denature a set of short sequence repeats in the replication origin in order that helicase will be able to bind and set up the replication fork.
    D. methylate a set of short sequence repeats in the replication origin in order to prevent multiple rounds of initiation events from occurring.
    E. generate the RNA primers required by DNA polymerase α.

10. The protein responsible for maintaining the separation of the DNA template strands after they have been unwound is

    A. DNA polymerase III.
    B. primase.
    C. DNA polymerase I.
    D. topoisomerase.
    E. single-strand binding protein.

11. Comparison of the proteins required for replication in *E. coli*, eukaryotic cells and the *E. coli* T4 virus suggests that

    A. the set of T4 enzymes is markedly different from either cellular one.
    B. the eukaryotic replication system has diverged greatly from that seen in bacteria and bacterial virus systems.
    C. similar enzymatic functions are required for replication in all three systems.
    D. the three DNA polymerases are markedly different with little conservation of either the amino acid sequence or the three dimensional structure.
    E. the major differences between the systems lie in the way discontinuous strand replication is accomplished.

12. The two *E. coli* polymerases that are responsible for most error-prone DNA repair are

    A. DNA polymerase I and III.
    B. DNA polymerase I and IV.
    C. DNA polymerase I and V.
    D. DNA polymerase II and III.
    E. DNA polymerase IV and V.

13. The function of primase in *E. coli* replication is to

    A. control the initiation of replication at the *oriC* locus.
    B. prime the replication of the ends of the linear chromosome.
    C. activate *de novo* initiation of replication by DNA polymerase III.
    D. generate short RNAs that DNA polymerase can use to start replication.
    E. form single-strand nicks to prime replication of both of the continuous and discontinuous strands.

14. Some mutations in the *E. coli* DnaA protein, which recognizes repeats in the replication origin, are suppressed by mutations in RNA polymerase. RNA polymerase most likely affects initiation of replication by

    A. producing the RNA primers required by DNA polymerase III.
    B. forming part of the replisome complex.
    C. changing the supercoiling of the origin region, making it easier to denature.
    D. binding to an adjacent promoter and forming a protein-protein interaction that helps the DnaA protein to bind to the origin.
    E. binding to the DnaA protein and substituting its helicase and polymerase functions for those of the DnaB and DnaG proteins.

15. After initiation of replication in *E. coli*, replication cannot reinitiate for about 15 minutes. During this time the *oriC* region remains hemimethylated and bound to the cell membrane. One protein that is required for delaying rereplication is _____.

    A. DnaB
    B. SeqA
    C. DNA polymerase III
    D. RecA
    E. Dam methylase

Answers:

A. Recall and short answer

1. $5' - 3'$ polymerization for DNA synthesis, $3' - 5'$ exonuclease for proofreading, and $5' - 3'$ exonuclease for RNA primer removal

2. helicase to separate the parental duplex strands, single-strand binding protein to maintain and protect the single-strand DNA, primase to synthesize the RNA primer, DNA polymerase III to accomplish most of the DNA synthesis, DNA polymerase I to remove the RNA primer nucleotides and replace them with deoxynucleotides, and DNA ligase to join the DNA fragments into a single polymer

3. the gene for DNA polymerase I due to the failure to remove the RNA primers or to replace the RNA nucleotides with deoxynucleotides, and the gene for DNA ligase due to the failure to join the fragments into a single polymer

4. Nicks in DNA are the primers for DNA synthesis during DNA repair and for the replication of some virus genomes.

5. The priming proteins have a deoxynucleotide covalently attached to the protein. The $3'$ OH group of the nucleotide serves as the primer.

6. The DnaA protein binds cooperatively to four 9 base bp repeats in the 245 bp OriC region as the first step of initiation of replication in *E. coli*.

7. DNA repeats that serve as the binding sites for proteins involved in initiating replication and A·T-rich regions at which separation of the DNA occurs.

8. The replicative polymerases are large, multisubunit proteins with high fidelity, while the repair polymerases are small, typically monomeric proteins with lower fidelity.

9. DNA polymerase α and DNA polymerase δ

10. licensing factors

11. the ORC complex, the Cdc6 protein, and the MCM licensing factors

12. One of the events during T4 infection is the degradation of the cellular DNA, therefore the genes for the host enzymes are destroyed and must be replaced with viral ones.

13. An Okazaki fragment is one of the DNA fragments made during discontinuous strand replication of the lagging strand. Initially it contains a 5 triphosphate group as part of a short RNA region, as nick translation by DNA polymerase I proceeds the triphosphate group and the RNA nucleotides are cleaved off.

14. Nick translation, $5 - 3$ exonuclease, $5 - 3$ polymerase.

15. *dna*B, quick-stop; *dna*A, slow-stop; *pol*C, quick-stop

B. Analytical and critical thinking

1. The most conserved structural domain is the catalytic domain (palm). Additional domains position the template strand in the active site (fingers) and bind the DNA as it exits the enzyme (thumb). Other domains provide the $3' - 5'$ exonuclease activity. Conservation of the sequence and/or three-dimensional shape of the enzyme may arise by descent from a common ancestor or by convergent evolution. Similarity of sequence or shape implies the presence of selective pressure that either maintains the ancestral form or favors mutations that bring DNA polymerases to a common structure. Since DNA replication is an essential function, descent from a common ancestor is more likely.

2. The genomic DNA provides the template, the oligonucleotides provide the primers with their $3'$ OH group, and the deoxynucleotide triphosphates are the small substrates required for the polymerization of the DNA.

3. A 245 bp region is predicted to have one GATC site. If each of the four possible base pairs occurs with equal frequency, then the likelihood of a specific base pair occurring at a site is 1/4. Since there are four base pairs in the GATC site, the probability of a GATC site occurring is $(1/4)^4$ or one site in every 256 bases on average. Therefore there are many more GATC sites than are predicted for a random DNA sequence of this length. Replication converts the fully methylated sites into hemimethylated sites that cannot serve as origins.

4. The major subcomplexes in DNA polymerase III are the clamp, clamp loader, and the core enzyme. The core enzyme catalyzes the synthesis of DNA and the proofreading of DNA. The clamp encircles the template DNA strand and slides along it like a bead on a string, keeping the DNA polymerase associated with the template strand. This increases the processivity of the enzyme. The clamp loader assembles the clamp on the DNA and mediates its interaction with the core enzyme. The core enzyme catalyzes the synthesis of DNA and the proofreading of DNA. In the replisome a single clamp loader complex, two clamps, two core enzymes, and two copies of the additional dimerization subunit τ form a 900 kD complex that replicates both the leading and the lagging strand.

5. The mutation may have arisen by replication slippage. When the replication fork stalls in the run of A·T base pairs due to damage in the template strand the polymerase disassociates from the DNA. After the damage is repaired reassociation of the polymerase with the DNA may occur. During this time the newly replicated strand and the template strand denature and then renature. If the renaturation is not precise in the run of AT base pairs, then the newly replicated strand will have more or fewer bases than are in the parental strand. If not corrected, this will lead to a frameshift mutation. In the example the mRNA will be shorter by one base and all the codons from the position of the mutation to the 3 end of the open reading frame will be incorrect. The polypeptide will have a correct amino terminus but will have incorrect amino acids at its carboxyl terminus. The polypeptide will also be a different length than the wild type because the stop codon of the open reading frame will be out of frame. The mutant polypeptide may be shorter or longer than normal. There may also be polar effects due to the frameshift mutation on the expression of downstream open reading frames in polycistronic mRNAs.

6. In typical replication forks replication of the leading strand is continuous from the initial priming event at the replication origin until replication terminates. Lagging strand replication requires many new primers to be used and therefore is discontinuous. The overall process is a

mixture of the two and therefore "semidiscontinuous". Replication of ϕX174 progeny strands requires only leading strand replication and therefore is continuous.

C. Extensions and applications

1. The error-prone polymerases allow cells to at least complete DNA replication on damaged templates. This gives a cell a chance to survive, which it otherwise will not have if it is not able to replicate its DNA. The error-prone enzymes are kept at low abundance until they are needed. PolIII and PolI do most of the replication of the cellular DNA. They are specifically adapted to initiate replication at the replication origin and to work together to accomplish discontinuous strand replication. Since PolII is capable of error-free replication restart, replication will be able to resume at most of the damaged sites before the error-prone PolV becomes available in large amounts. This means that the damaged region is correctly replicated and the damage can be repaired by photoreactivation or excision repair processes before the next round of replication. Therefore the number of mutations will be much lower. The active sites of the enzymes are all different. The amino acids at the active site are not identical and therefore there are three-dimensional differences that allow some damaged templates to be used by one enzyme and not by others. Likewise the three-dimensional features of the sites allow entry and use of specific nucleotides at the damaged sites. Having all the enzymes work with the same processivity clamp allows much easier access to the template. One can imagine an efficient transfer of the clamp from PolIII to PolI during discontinuous strand replication as well as transfer of the clamp to one of the repair polymerases in response to a damaged template. Whenever PolIII is incapable of using the template, the clamp is likely to be passed to another enzyme. In undamaged cells the enzyme is likely to be PolI. In damaged cells the enzymes will include a mix of PolI, PolII, PolIV and PolV. The one that is used will depend in part on the abundance of the enzyme in the cell. If one is chosen that is unable to use the damaged template, then it in turn may be exchanged for one of the other polymerases.

2. The function of the ORC complex is to bind to and identify the replication origins for the licensing factors. The ORC complex is bound throughout the cell cycle. Similar proteins are known to be present in many eukaryotes, including man. Licensing factors are thought to gain entry into the nucleus when the nuclear membrane disintegrates during mitosis. Phosphate groups are added to serine, tyrosine, or threonine. Addition of the negatively-charged phosphate group alters the three-dimensional shape of the protein. This may either activate or inactivate the function of the protein. Since both are needed in order to license replication, having multiple and differently regulated mechanisms controlling the activity of the licensing factors provides better control over the process. With a single type of control a breakdown in the regulation through a mutation may lead to unwanted rounds of replication. Reinitiation of replication is prevented at least in part by the lack of the hCtd1 protein, which is removed by proteolytic degradation.

D. Standardized test format

1. B
2. A
3. E
4. C
5. D
6. A
7. D
8. A
9. B
10. E
11. C
12. E
13. D
14. C
15. B

Chapter 15

Recombination and repair

General Concepts	Section	Related Questions
Recombination involves breakage and reunion	15.1-15.4	A1, A4, A13
Recombination in bacteria	15.8-15.10	A9, B1, B6, D9
Gene conversion involves mismatch repair	15.11	A10, B4
Supercoiling affects the topology of DNA	15.12	A14, B2, D1, D4
Specialized recombination involves specific sites	15.16-15.19	A1, A5, D5, D8
DNA damage can lead to mutations	15.20	A6, C1, B5, D2, D6, D12
DNA repair corrects damaged genomes	15.20-15.22	A7, A8, A15, B5, D7, D15
Error-prone repair causes mutations	15.23, 15.27	B1, C2, D14
Mismatch repair can be strand specific	15.24	A10, B3, D13
Recombination repair aids replication	15.25-15.26	B1
Defective repair causes human diseases	15.28-15.29	A11, A12

Processes/structures		
Synaptonemal complex	15.5-15.7	A2, D11
Heteroduplex formation	15.3	A3, A9
Single-strand exchange	15.8-15.9	B1, B6
Resolution of joint molecules	15.10	A4, A9, D9
Topoisomerase action	15.13-15.15	A14, B2, D1, D4
Mechanism of site-specific recombination	15.17-15.19	A1, A5
Photoreactivation	15.20, 15.22	A8, D3
Excision repair	15.20-15.22	A7, A8, A15, B5, D7, D15
Mismatch repair	15.20, 15.24	B3, B4, D13
Error-prone repair	15.23, 15.27	B1, C2
Recombination repair	15.25-15.26	B1, B6
Nonhomologous end-joining	15.29	A12

Examples, Applications and Techniques		
Holliday model	15.2-15.3	A3
Double-strand break model	15.4, 15.6	D10, D11
E. coli recombination enzymes	15.8-15.10	A9, B1, B6, D9
Type I and Type II topoisomerases	15.13-15.15	A14, B2, D1, D4
Lambda site-specific recombination	15.16-15.19	A5, D5
Eukaryotic repair enzymes	15.28-15.29	A11, A12

Questions:

A. Recall and short answer

1. What are the differences between homologous recombination and site-specific recombination?

2. The _____ is the morphological structure formed by pairing two homologous chromosomes during meiosis.

3. What is a heteroduplex and how is it formed?

4. The two types of DNA molecules produced by resolution of recombining chromosomes are _____ in which the DNA molecules contain a mixture of duplex regions derived from each parental duplex plus some heteroduplex DNA and _____ in which the parental duplexes remain intact outside of a heteroduplex region.

5. The DNA sites required for integration of lambda DNA into the bacterial chromosome are the viral _____ and the bacterial _____ sites.

6. Deamination of cytosine in DNA generates _____, while deamination of the modified form of cytosine 5-methylcytosine generates _____.

7. Which is the major DNA polymerase associated with excision repair in *E. coli*?

8. The two processes that actually repair pyrimidine dimers in *E. coli* DNA are _____ and _____.

9. During branch migration in *E. coli* _____ recognizes the recombination joint, the _____ catalyzes branch migration, and _____ cleaves the DNA strands to resolve the joint molecule.

10. _____ is the nonreciprocal transfer of information from one DNA molecule to another. It depends on the presence of _____.

11. One of the effects seen in people with defective DNA repair is _____.

12. The ligation of the blunt ends of two DNA duplexes is carried out by the enzymes of the _____ pathway.

13. Many RNA viruses have the ability to recombine by a copy choice mechanism in which _____.

14. The enzymes that control supercoiling are _____. Those of _____ add or remove one supercoil each time they act, while those of _____ add or remove two supercoils each time they act.

15. Enzymes that break the covalent bond that links a base to the deoxyribose group in DNA are called _____.

B. Analytical and critical thinking

1. What are the roles of the RecA protein in *E. coli* cells? What conditions are needed for it to function in each role?

2. In a typical molecular genetics experiment a circular plasmid DNA serves as a cloning vector. To obtain a sample of the recombinant DNA the plasmids that carry inserts must be isolated in pure form from the bacterial cells. These preparations typically contain three topological forms of the plasmid DNA. These are termed covalently closed circular DNA, open circular DNA and linear DNA. Which of these three forms can be supercoiled? How many and what types of nicks or double-strand breaks are present in each of the three forms? Gel electrophoresis can be used to separate the three forms of plasmid DNA, which form will have the greatest mobility, i.e., can pass through the gel easiest and why? In *E. coli* there is about one supercoil per 200 bp of duplex DNA; how many supercoils are predicted to be present in a 3000 bp cloning vector?

3. The bases found in DNA and RNA can transiently take on alternative conformations. In these tautomer conformations they pair incorrectly, for example the tautomeric form of T will pair with G and not with A. If a tautomeric T is incorporated into replicating DNA and then changes back to the normal conformation, what does the DNA duplex look like? How will this duplex differ from that generated when a 5-methylcytosine in duplex DNA undergoes a deamination event?

4. Gene conversion is often studied using *Ascomycetes* spores where spore color phenotypes are easily followed. In this species the four products of meiosis undergo an additional mitotic division to give a total of eight spores that are aligned linearly in the ascus. Most asci have spores aligned in a 4 wild type:4 mutant pattern. Others are aligned in 3:5 or 2:6 patterns or even patterns such as 2:2:2:2. How are the 3 wild type:5 mutant patterns generated? How are the 2 wild type:2 mutant:2 wild type:2 mutant patterns generated?

5. Nitrous acid is a mutagen that increases deamination of cytosines in duplex DNA. What is the product produced by the deamination of cytosine? If this damage is not repaired what type of mutation will occur? Describe a repair process that specifically repairs this deamination product.

6. What is the substrate for the action of the RecBCD nuclease in *E. coli* and what is the structure of the product made after nuclease action? How does this product relate to the substrate required by the RecA protein for its action?

C. Extensions and applications

1. Mutations arise as a consequence of DNA damage but most damage is repaired. Therefore both processes influence mutation rates. Some types of damage are efficiently repaired and so no mutations result. Other types of damage, although rare, are not as efficiently repaired, and therefore cause many mutations. Nachman and Crowell (2000 *Genetics* 156:297-304) estimated the mutation rate in humans based on sequence comparisons of 18 pseudogenes in humans and chimpanzees. An average mutation rate of about 2.5×10^{-8} mutations per nucleotide per generation was determined. This is equivalent to about 175 mutations per diploid genome per generation! They also assessed the mutation rate for several different types of mutations. The rate of mutation at CpG dinucleotides was about 10-fold greater than at other types of sites and the rate of base substitution mutations was also about 10-fold greater than that of insertion/deletion

mutations. Transition mutations occurred 2 to 4 fold more frequently than transversion mutations. A total of 131 transition, 52 transversion and 16 insertion/deletion mutations were observed. About one fifth of all the mutations were transition mutations at CpG dinucleotides.

Transition mutations are base substitution mutations in which, for example, one pyrimidine-purine base pair is replaced by the other possible pyrimidine-purine base pair. Transversion mutations are base substitution mutations in which, for example, a pyrimidine-purine pair is replaced by a purine-pyrimidine pair. How many different types of transitions and transversions are possible? For any specific base pair in DNA, how many of these possibilities may occur? Transition mutations outnumbered transversion mutations 131 to 52 among the observed mutations; is this consistent with the number of possible types of transition and transversion mutations? If the observed numbers are not consistent, then suggest reasons why the numbers do not agree. Why do mutations occur so frequently at CpG dinucleotides; are the nucleotides in these sites more likely to be damaged or less likely to be repaired?

2. Adaptive mutations occur when microorganisms are exposed to stressful but not lethal selection. Under these conditions mutations that relieve the selection may arise. Both beneficial and nonbeneficial mutations are now known to take place under these conditions. The mutations are not targeted only to genes that produce products that directly relate to the selective condition and its relief. Much of the work on adaptive mutation has used the *E. coli* strain FC40. This strain cannot use lactose but produces mutants that can use lactose when this sugar is the only available carbon source. Most adaptive mutations arise through recombination mechanisms and require DNA nicks generated by conjugation proteins encoded on the F plasmid. Almost all adaptive mutations in the *lacZ* gene of FC40 are –1 bp frameshifts. During selection about 0.1 % of the FC40 cells undergo a period of greatly increased mutation rate, hypermutation. These cells have a mutation rate 200-fold higher than normal. About 10 % of all the Lac$^+$ mutations arise in these cells and the remaining 90 % in cells that are not in the hypermutation state using the recombination-dependent mechanism. Tompkins and coworkers (2003, Journal of Bacteriology 185:3469-3472) have investigated the role of error-prone polymerase PolIV in the hypermutation phenomenon. Both PolIV and PolV are induced by the SOS response and are highly error-prone, even when replicating undamaged DNA. PolV can carry out translesion synthesis on damaged templates; PolIV can use primers that are not correctly base-paired at their 3′ ends to the template strand. Experimental removal of PolV activity during adaptive mutation previously was shown to have no effect on the production of Lac$^+$ cells. In the absence of PolIV the rate of adaptive mutation to the Lac$^+$ phenotype fell 3 to 5 fold.

How are the genes in the SOS regulon normally induced? What does this suggest about the DNA in the cells in which hypermutation occurs during nonlethal selective pressure? Is the hypermutation state a permanent change to the cell or a transitory state? Is a –1 bp frameshift consistent with the known activity of PolIV? What do you predict will happen to the rate of nonselected mutations in cells deficient for PolIV?

D. Standardized test format

1. The major enzyme responsible for introducing negative supercoils into DNA during replication in *E. coli* is

    A. topoisomerase I.
    B. helicase.
    C. topoisomerase IV.
    D. topoisomerase III.
    E. gyrase.

2. Since genes, mRNAs and proteins are colinear in bacteria, mutations in an *E. coli* gene that change the sequence near the start codon of a gene will change

    A. the nucleotide sequence near the 3′ end of the open reading frame in the mRNA and an amino acid near the carboxyl terminus of the protein.
    B. the nucleotide sequence near the 5′ end of the open reading frame in the mRNA and an amino acid near the carboxyl terminus of the protein.
    C. the nucleotide sequence near the 3′ end of the open reading frame in the mRNA and an amino acid near the amino terminus of the protein.
    D. the nucleotide sequence near the 5′ end of the open reading frame in the mRNA and an amino acid near the amino terminus of the protein.
    E. the nucleotide sequence near the 3′ end of the open reading frame in the mRNA but will not change an amino acid in the protein.

3. In *E. coli* pyrimidine dimers can be repaired by photoreactivation. This repair process involves

    A. excision of the dimer by the UVR enzymes.
    B. breaking two covalent bonds in the dimer using energy from ultraviolet light.
    C. breaking two covalent bonds in the dimer using energy from visible light.
    D. gap filling opposite the dimer using recombination enzymes.
    E. replication past the dimer by an error-prone DNA polymerase.

4. In a typical topoisomerase I action

    A. a single-strand break is made, a single DNA strand is passed through the gap, and the break is resealed.
    B. a double-strand break is made, a double-strand duplex is passed through the gap, and the break is resealed.
    C. a single-strand break is made, a double-strand duplex is passed through the gap, and the break is resealed.
    D. a double-strand break is made, a single DNA strand is passed through the gap, and the break is resealed.
    E. a double-strand break is made, the ends of the DNA are rotated 360 degrees relative to one another, and the break is resealed.

5. Integration of the lambda virus into the *E. coli* chromosome occurs by

    A. staggered cleavage of viral DNA strands, formation of a blunt-ended double-strand break in the bacterial DNA, joining of the ends of the two DNAs, gap filling by DNA polymerase, and ligation of the last phosphodiester bond.

    B. staggered cleavage of the viral and the bacterial DNAs, exchange of ends to generate a short heteroduplex region, and resealing.

    C. double-strand cleavage of the cellular DNA target site followed by double-strand break repair using the viral DNA as the donor.

    D. blunt-ended double-strand breakage in both DNAs followed by rejoining with no heteroduplex formation.

    E. double-strand cleavage of cellular DNA, annealing of a 3′ end in cellular DNA to a site in the viral DNA, replication across the viral genome, followed by another strand cleavage, and ligation of the last phosphodiester bond.

6. The major damage induced by ultraviolet light in DNA is

    A. depurination of guanine nucleotides.

    B. alkylation of adenine.

    C. deamination of cytosine.

    D. covalent bond crosslinks between the two strands in the duplex.

    E. pyrimidine dimers.

7. In *E. coli* excision repair of pyrimidine dimers which protein recognizes the damaged DNA; which protein removes the damaged DNA and which proteins replace the damaged DNA?

    A. RecA, RecBCD, and a combination of DNA polymerase I and DNA ligase.

    B. DnaA, RecBCD, and a combination of DNA polymerase I and DNA ligase.

    C. RecA, RecBCD, and a combination of DNA polymerase III and DNA ligase.

    D. UvrAB, UvrBC and UvrD, and a combination of DNA polymerase I and DNA ligase.

    E. UvrAB, UvrBC and UvrD, and a combination of DNA polymerase III and DNA ligase.

8. Bacterial plasmid maintenance systems sometimes include proteins that catalyze recombination at a specific site on the plasmid DNA. This converts dimer DNA molecules into monomers. Why is this important to the plasmids?

    A. Dimers are twice as big as monomers and will be less efficiently replicated.

    B. Monomeric forms are more easily integrated into the host cell chromosome.

    C. The dimers are too big to be packaged in the plasmid capsids, which are just big enough to carry the monomeric form to new cells.

    D. Monomeric DNA is less prone to DNA damage arising from replication errors.

    E. Plasmids are randomly segregated to daughter cells, having monomeric genomes increases the likelihood that each cell acquires a copy of the plasmid.

9. The *E. coli* endonuclease that specifically recognizes Holliday junctions and cuts the DNA strands in the recombination joint is

   A. RuvC protein.
   B. RecA protein.
   C. RuvB protein.
   D. RecBCD protein.
   E. RecG protein.

10. The function of the Spo11 protein in *Saccharomyces cerevisiae* recombination is

   A. to catalyze branch migration.
   B. to carry out the initial single-strand assimilation.
   C. to generate a double-strand break to initiate recombination.
   D. to form single-strand DNA by exonuclease action.
   E. to cleave strands to separate the two chromosomes at the end of recombination.

11. Genetic recombination during meiosis is initiated by a

   A. pair of single-strand nicks at identical sites, one within each parental duplex.
   B. single-strand nick in only one of the two parental duplexes.
   C. pair of single-strand nicks at different sites, one within each parental duplex.
   D. double-strand break in only one of the two parental duplexes.
   E. pair of double-strand breaks at identical sites within each parental duplex.

12. Microsatellites are tandem arrays of short DNA repeats. They are found at many places in genomes and are very useful in DNA fingerprinting analyses due to the extensive genetic variability in the number of repeat units present in different individuals. Mechanisms for generating the genetic variability include replication slippage and

   A. site specific recombination.
   B. unequal crossing-over during homologous recombination.
   C. mismatch repair
   D. topoisomerase action.
   E. excision repair.

13. During mismatch repair of replication errors in *E. coli*

   A. part of the newly replicated, nonmethylated DNA strand is removed and replaced by DNA polymerase I.
   B. part of the newly replicated, methylated DNA strand is removed and replaced by DNA polymerase I.
   C. the incorrect base is flipped out of the helix and removed by a glycosylase, the AP site is excised and the region replaced by DNA polymerase I.
   D. the repair enzymes cannot distinguish the parental strand from the newly replicated strand and therefore about half of the mistakes lead to mutations.
   E. the region of the strand containing the mistake is removed and replaced by transfer of a single-strand region from another DNA duplex.

14. The *E. coli* SOS repair system involves

   A. inhibition of the proofreading activity of DNA polymerase I.
   B. replacement of DNA polymerase III in the replication fork by DNA polymerase IV or DNA polymerase V.
   C. induction of lambda virus repair enzymes.
   D. decreased replication slippage by DNA polymerase III.
   E. inhibition of the proofreading activity of DNA polymerase III.

15. Uracil in DNA is

   A. converted *in situ* directly to thymine by a repair process.
   B. not removed from DNA because it pairs normally with A.
   C. removed by a glycosylase-based form of excision repair.
   D. removed by a repair system that begins with an endonuclease binding the uracil-containing strand and then nicking it on either side of the uracil.
   E. converted *in situ* directly to cytosine by a repair process.

Answers:

A. Recall and short answer

1. Homologous recombination can occur at any position in a DNA duplex and requires long regions of homology, while site-specific recombination can only occur at particular sequences and utilizes short regions of homology.

2. synaptonemal complex

3. Heteroduplexes are duplex DNA molecules in which the two strands come from different parental DNA duplexes. They are formed by strand exchange and branch migration.

4. splice recombinants, patch recombinants

5. *att*P, *att*B

6. uracil, thymine

7. DNA polymerase I

8. photoreactivation, excision repair

9. the RuvA protein, the RuvB helicase, the RuvC protein

10. gene conversion, mismatched bases in heteroduplex regions

11. an increased likelihood for developing cancer

12. NHEJ (nonhomologous end joining)

13. the viral RNA polymerase switches from one template to another while making the new RNA strand

14. topoisomerases, Type I, Type II

15. glycosylases.

B. Analytical and critical thinking

1. RecA protein has two roles. The first is to catalyze the assimilation of single-strand DNA into a homologous DNA duplex. It does this by displacement from the duplex of the strand that has the same, or very similar, sequence as that of the single-strand DNA. This process requires at least a single-strand region on one DNA duplex, a free 3′ end, and a complementary region between the two DNA molecules that includes both the single-strand region and the 3′ end. The second role is to activate the autocleavage of the LexA repressor during induction of the SOS system. The inducing signal used by RecA is not known but may be a small molecule released from the heavily damaged DNA or a structure within the damaged DNA.

2. Only covalently closed circular DNA with no nicks or strand breaks can be supercoiled. Open circular DNA is relaxed and at least has no supercoils, typically it has one or more single-strand nicks in the DNA. The presence of even one nick will cause the DNA to lose all supercoiling. Linear DNA has a double-strand break, the ends of the break may have short single-strand tails if the breakage was asymmetric. Supercoiled DNA will move the furthest because it has a more compact shape. There may be up to 15 supercoils on a 3000 bp plasmid.

3. It has a mismatched TG pair. Since the tautomeric T is in a newly replicated strand it will be in a hemimethylated DNA duplex. This is different from the fully methylated duplex that would be the case after the deamination of a 5-methylcytosine.

4. The 3:5 patterns must involve formation of heteroduplexes that cross the spore color gene. Mismatch repair on one heteroduplex converts one wild type strand to the mutant sequence. If the other heteroduplex does not undergo repair, then the mitotic replication gives the 3:5 pattern. For the 2:2:2:2 pattern there is 4 of each sort of product. These also may derive from heteroduplexes and mismatch repair. However, a more likely explanation is that a single recombination event has taken place between the centromere and the spore color gene.

5. The deamination product of cytosine is uracil. If not repaired there will be a GC to AT single base change. These are also described as base substitution mutations, point mutations, or transition mutations. A glycosylase flips the uracil out of the helix and cleaves the glycosidic bond. The region of the DNA strand missing the base is removed and replaced by an excision repair process.

6. The enzyme binds to any fragment of DNA with a double-strand break. These may be products from stalled replication forks or may be DNA fragments acquired by bacterial genetic exchange such as transduction and/or restriction enzyme digestion of transduced DNA. The enzyme moves along the fragment from the double-strand break unwinding the strands and degrading one strand of the DNA as it goes until in encounters a chi sequence, GCTGGTGG. It cleaves the strand it was degrading just downstream of the 3' side of the chi sequence. After this the helicase activity continues to denature the DNA duplex but further degradation of single-strand DNA does not occur. This generates a single-strand DNA with a 3′ OH group that is the substrate required by the RecA protein for its action.

C. Extensions and applications

1. There are four possible types of transition mutations and eight possible types of transversion mutations, each specific base pair can mutate by only one sort of transition mutation and two sorts of transversion mutations. Therefore there are twice as many possibilities for transversion mutations as for transition mutations but transition mutations occurred at a much higher rate than predicted by this simple sort of calculation. Mutation rates reflect both the occurrence of DNA damage and the repair of that damage. Damage that leads to transition mutations occurs much more often than damage that leads to transversion mutations. Errors by DNA polymerase for example are predominantly the substitution of one pyrimidine or one purine for a nucleotide of the same type, therefore generating transition mutations if not repaired. Likewise damaged bases in DNA that are able to pair with one of the normal bases are more likely to cause transition mutations than transversion mutations. Spontaneous deamination of cytosine and adenine likewise typically lead to transition mutations if the damaged base is not repaired. Mutations occur frequently at CpG dinucleotides not because these sites are more easily damaged but because they are less easily repaired. The cytosines in the CpG dinucleotides are actually 5-

methylcytosines, deamination of 5-methylcytosine results in thymine. Cells are less able to correctly repair TG mismatches than they are able to repair UG mismatches. In addition to mismatch repair systems, cells have specific repair pathways to recognize and remove uracil from DNA.

2. Heavily damaged DNA is normally necessary to induce the SOS regulon. Either a product released from the damaged DNA or a part of the damaged DNA itself provides a signal recognized by the RecA protein. The RecA protein then interacts with the LexA repressor protein to induce the self-cleavage of the LexA protein. The LexA cleavage products are no longer functional as repressors and the genes of the regulon are induced. These genes include the PolIV and the PolV DNA polymerases. This suggests that some of the cells under stressful environmental conditions may have damaged DNA, alternatively the molecule used by RecA to recognize DNA damage is somehow produced or its action mimicked by some other means by the selective pressure. Hypermutation cannot be a permanent state. If it is not transitory, then the cell will eventually suffer a lethal mutation. A –1 bp frameshift is consistent with the known mode of action of PolIV, since this polymerase is able to synthesize DNA under conditions in which a perfect primer-template pairing is not required. In cells lacking PolIV activity the rate of the nonselective mutations elsewhere in the genome also decreased. The rate at which the nonselected mutations occurred fell seven-fold in the experiments reported by Tompkins and coworkers.

D. Standardized test format

1. E
2. D
3. C
4. A
5. B
6. E
7. D
8. E
9. A
10. C
11. D
12. B
13. A
14. B
15. C

Chapter 16

Transposons

General Concepts	Section	Related Questions
Transposons are mobile genetic elements	16.1-16.3	A1, C1, D1, D4, D5
IS elements are simple transposons	16.2	A2, A6, A7, B1
Transposons may carry selective traits	16.3	B1, C1, D7, D13
Transposition	16.6-16.9	A3, A4, A9, B5, B6, D8
Transposition frequency is controlled	16.10	A15, B2, B6, D14
Autonomous and nonautonomous elements	16.12-16.13	B3, D10
Transposable elements affect gene expression	16.13	A1, D9
Transposition causes hybrid dysgenesis	16.14-16.15	A12, B4, D11
Processes/structures		
Transposons end in inverted repeats	16.2	B1, D2
Target site direct repeats flank transposons	16.2	B5, B6, D2, D6
Conservative transposition	16.4	D8
Nonreplicative transposition	16.4	A8
Replicative transposition	16.4, 16.7	A8, A13, D8, D15
Transposons mediate DNA rearrangement	16.5	A5-A7, A10, D3
Resolution of cointegrates	16.8-16.9	A13, D12
Chromosome breakage	16.11	A8, A10,
Hybrid dysgenesis	16.14-16.15	A12, B4, D11
Examples, Applications and Techniques		
IS elements	16.2	A2, A6, A7, B1
Composite transposons	16.3	B1, D7, D13
Mu virus	16.6	A9
Tn3	16.9	A13, B1, D12
Tn10	16.10	A15, B2
Ac/Ds elements	16.11	A10, A14, B3, C2, D10
Spm elements	16.13	A11, B3, D9, D10
P elements	16.14-1615	A12, B4, D11

Questions:

A. Recall and short answer

1. Integration of transposons into genes most commonly leads to _____.

2. The simplest type of bacterial transposon is _____.

3. The sequence into which a transposon moves is termed its _____ and the site from which it moves is the _____.

4. The protein responsible for mediating the movement of a transposable genetic element is its _____.

5. Recombination events between two transposable elements on the same DNA molecule can lead to _____ and _____.

6. How can the presence of IS elements mediate the integration of a circular plasmid DNA such as the F plasmid into a cellular chromosome?

7. Removal of IS elements from a DNA molecule may involve _____ in which a recombination event occurs between the target site direct repeats flanking the element or by _____ in which the recombination event leaves a remnant of the IS element in the DNA.

8. Replicative transposition generates _____ by joining the donor and target DNA. Nonreplicative transposition generates _____ in the donor DNA after movement of the transposon.

9. How is the MuA transposase involved in each of the three stages of Mu virus transposition?

10. Chromosome breakage is associated with transposition of the Ac/Ds elements in maize. During the classic breakage-fusion-bridge cycle DNA molecules with no centromere and ones with two centromeres are formed. These are termed _____ and _____ respectively.

11. Heritable but reversible changes in the expression pattern and mobility of transposable elements are examples of _____.

12. _____ is the correlated coappearance of mutations, chromosome aberrations, and sterility arising from the movement of transposable genetic elements in *Drosophila*.

13. What are the functions of the TnpA and the TnpR proteins of the Tn3 transposon?

14. The production of sectors of different colors on maize seeds is an example of _____ due to the movement of transposable genetic elements.

15. How does pairing of the IN and OUT RNAs produced from the Tn10 transposon decrease expression of the transposase?

B. Analytical and critical thinking

1. Compare the structures of an IS element, of a composite transposon and of a transposon of the TnA family. What types of genes and sequence features are present in each type of transposable element?

2. How is DNA methylation used to regulate Tn10 transposition?

3. What is the relationship of nonautonomous transposable genetic elements to autonomous ones?

4. What is the difference between the P element mRNA that encodes the repressor protein and that which encodes the transposase protein?

5. How are the direct repeats of the target site generated?

6. Not all transposons integrate randomly, some have preferred target sites. How are preferred target sites identified?

C. Extensions and applications

1. Transposons in bacteria are not limited to IS elements and composite transposons. More complicated transposons include the elements of the TnA family and phages such as Mu. Other elements termed conjugative transposons exist. As suggested by their name these transposons carry genes required for conjugation as well as transposition. Wang and coworkers (2003 *Applied and Environmental Microbiology* 69:4595-4603) describe a new conjugative transposon from an isolate of a *Bacteroides* species that includes an erythromycin resistance gene. The CTnGERM1 conjugative transposon is about 75 kb in size with a G + C content significantly lower than that of the rest of the *Bacteroides* chromosome. The *erm*G gene is only 26.8 % G + C and flanking genes in the transposon are about 35 % G + C compared to a chromosomal 42 % G + C content for the gram-negative *Bacteroides* species. The *erm*G gene on the transposon is 99 % identical to a gene previously found in a gram-positive bacterium, *Bacillus sphaericus*. Conjugation is thought to be the method involved in broad host range genetic transfer among bacterial species. Initial transfer of single-strand DNA during conjugation helps the DNA to escape the restriction modification systems that are targeted towards double-strand DNA in the recipient cell.

What types of genes and sequence features are predicted to be present on a conjugative transposon? What selective advantage may conjugative transposons provide to cells? How would one show that it integrates into chromosomal DNA rather than existing as an extrachromosomal episome? What appears to be the origin of the CTnGERM1 transposon?

2. Although the maize Ac/Ds transposons were discovered decades ago, they have not been used extensively in gene targeting experiments. Singh and coworkers (2003 *The Plant Cell* 15:874-884) are developing approaches to use Ac for regional mutagenesis throughout the maize genome. This requires the development of nearly isogenic lines with single elements distributed at well-defined positions about 10 or 20 centimorgans apart throughout the genome. Ac displays a propensity for transposition to sites nearby to its starting location, hence the need for numerous isogenic lines. Ac also has a relatively low transposition frequency. However, Ac transposition events can be easily selected based on kernel spotting patterns. Heterozygous F1 plants derived from the transposition seeds can in turn be allowed to self-fertilize to produce F2 progeny among

which will be individuals homozygous for mutations resulting from the transposition event. These can then be scored to identify interesting mutations.

Propose mechanisms to account for the propensity of a transposon to move to nearby rather than distant target sites. What are the advantages of using the Ac system rather than one in which the target sites for transposition are more dispersed across the genome? What is an advantage of a system in which the transposition frequency is relatively low? Some of the mutant genes generated in these experiments retained partial function, how is that possible, don't transposition events lead to insertional gene inactivation? Most chromosomes resulting from the transposition events have the Ac element shifted from its original location to a nearby site. Presumably the movement of the element did not leave a double-strand break that could not be repaired at the original location of the element. What are three mechanisms that may be used to repair double-strand breaks left after transposition events?

D. Standardized test format

1. In bacterial genomes transposable elements are typically

    A. found as single copy DNA sequences in the bacterial chromosome.
    B. found as highly repetitive DNA sequences present in 1000's of copies.
    C. found as moderately repetitive sequences present in a few copies of each sort.
    D. not present due to selection to maintain a small genome size.
    E. found only on plasmid DNA not the bacterial chromosome itself.

2. After a bacterial IS element has moved into a new region of the genome, what will the DNA sequence look like, if P = part of the original target DNA sequence, IR = an inverted repeat, DR = a direct repeat, and T = the transposase gene of the element.

    A. 5′ P - IR - DR - T - IR - DR - P 3′
    B. 5′ P - DR - IR - T - IR - DR - P 3′
    C. 5′ P - IR - DR - T - DR - IR - P 3′
    D. 5′ P - DR - IR - T - DR - IR - P 3′
    E. 5′ P - DR - DR - T - IR - IR - P 3′

3. Transposable genetic elements are involved in generating DNA rearrangements because they

    A. inactivate the genes into which they insert.
    B. make short direct repeats at their insertion sites.
    C. often carry antibiotic resistance genes.
    D. provide repeated sequences at which unequal crossing-over may occur.
    E. can be carried on plasmid DNAs.

4. Most mutations that are associated with transposable genetic elements involve

A. deletion of a portion of the genome by recombination between two copies of a transposable element.

B. duplication of a portion of the genome by recombination between two copies of a transposable element.

C. inversion of a portion of the genome by recombination between two copies of a transposable element.

D. precise excision of a transposable element by recombination between the target site repeats flanking the element.

E. disruption of a gene by insertion of a new copy of a transposable element.

5. Which of the following is NOT true?

A. Different maize plants have different numbers of Ac/Ds elements.

B. Fruit fly individuals may or may not have copies of the P element.

C. Transposons of the same type occur at different sites in different individuals.

D. Transposons are identically distributed in all individuals in a species.

E. Transposons of several different types are present in a typical bacterial cell.

6. The direct repeats of the target site that flank transposable elements

A. arise from the staggered strand cleavage of the target site by the transposase.

B. reflect the preferred integration of transposable elements into repetitive DNA.

C. are identical in sequence for all copies of a specific transposon.

D. are carried with the element as it moves from site to site.

E. are used by the transposase to identify the ends of the transposon.

7. A major force supporting the continued existence of a composite transposon as a single unit rather than splitting it into independent IS elements is that

A. the mechanism of transposition ensures that larger DNA fragments are more efficiently bound by the transposase than are smaller DNA fragments.

B. the mechanism of transposition ensures that smaller DNA fragments are more efficiently bound by the transposase than are larger DNA fragments.

C. there is selection for the additional genetic traits carried by the composite transposon.

D. there is selection against the additional genetic traits carried by the composite transposon.

E. to form a functional transposase subunits from both of the IS elements must work together in a multimeric protein.

8. What is the difference between conservative and replicative transposition?

    A. In conservative transposition the number of copies of the transposon stays the same, but in replicative transposition the number of copies decreases by one.
    B. In conservative transposition the number of copies of the transposon stays the same, but in replicative transposition the number of copies increases by one.
    C. In conservative transposition the donor site is rejoined after movement of the transposon, but in replicative transposition the site is left with a double-strand break.
    D. In conservative transposition the number of copies of the transposon stays the same and there is a double-strand break at the donor site, but in replicative transposition the number of copies increases by one and there is no strand break at the donor site.
    E. In conservative transposition insertion of the transposon into genes does not occur, but in replicative transposition insertional gene inactivation occurs frequently.

9. The mechanism by which Spm elements suppress gene expression is that the

    A. insertion of an Spm element within an intron blocks splicing of the intron from the mRNA.
    B. binding of the Spm TnpA protein to the element blocks the initiation of transcription of the gene by RNA polymerase.
    C. insertion of the Spm element increases the methylation of the promoter of the gene thus decreasing the level of gene expression.
    D. insertion of the Spm element decreases the methylation of the promoter of the gene thus decreasing the level of gene expression.
    E. binding of the Spm TnpA protein to the element blocks transcription of the gene in which the element is located.

10. In the absence of an autonomous element a nonautonomous element will

    A. not be able to transpose.
    B. cause single-strand breaks whenever it transposes.
    C. not be able to inactivate the expression of the gene in which it resides.
    D. be recognized and deleted by the cellular DNA repair mechanisms.
    E. undergo a change of phase to regain the ability to transpose on its own.

11. Hybrid dysgenesis based on the P element of *Drosophila* occurs when

    A. a P male mates with a P female.
    B. an M male mates with a P female.
    C. a P male mates with an M female.
    D. an M male mates with either a P female or an M female.
    E. a P male mates with either an M female or a P female.

12. Resolution of Tn3 cointegrates

    A. is dependent on the cellular recombination enzymes to carry out the recombination event.

    B. is dependent on the presence of a lambda virus prophage to supply the enzymes required to carry out site-specific recombination.

    C. requires the TnpR protein to complex with the cellular RecA protein in a multimeric complex that will carry out site-specific recombination.

    D. generates a single fused DNA molecule from two independent starting DNAs.

    E. involves the action of TnpR to bind to the res sites, to generate strand breaks at the res site, to catalyze strand exchange, and to rejoin the strands.

13. The relationship between the occurrence of bacteria resistant to multiple antibiotics and transposable genetic elements is that these bacteria

    A. typically contain a plasmid on which there are several composite transposons bearing different drug resistance genes.

    B. activate the expression of drug resistance genes within their genomes by changing the methylation patterns of sequences present in the transposons.

    C. inhibit the expression of drug susceptibility genes within their genomes through transposon-mediated insertional gene inactivation.

    D. repress the expression of drug susceptibility genes by the interaction of transposase with the promoters of the susceptibility genes.

    E. have heavily rearranged genomes as a consequence of the presence of large numbers of IS elements.

14. In plant genomes the major mechanism to control transposition frequencies of transposons such as the Ac/Ds or the Spm element is

    A. through the action of antisense RNAs that bind to the transposase mRNA.

    B. to target the transposase for degradation by a protein modification.

    C. by activation of an exonuclease that specifically targets transposase mRNA.

    D. to change the methylation pattern of the promoters of the transposase genes to decrease expression of the transposase.

    E. to express a protein that interacts with the transposase protein, thereby not allowing it to bind to the transposon to initiate transposition.

15. The term "replicative transposition" describes the fact that

    A. the transposon only moves immediately after the replication fork has replicated the transposon.

    B. transposition requires a second copy of the transposon to be generated during the transposition event by replication across the original transposon sequence.

    C. the transposon can only transpose into the hemimethylated form of DNA produced by replication.

    D. the transposon only moves in the germ cells of the animals, a repressor of transposition is present in somatic cells.

    E. replication of the cellular chromosome is necessary for the cell to be able to use recombination to repair the double-strand breaks generated by the transposition events.

Answers:

A. Recall and short answer

1. insertional gene inactivation

2. an insertion sequence element (IS element)

3. new target site, donor site

4. transposase

5. deletions, inversions

6. A single homologous recombination event between a copy of the transposon on the plasmid DNA and a copy of the transposon on the chromosomal DNA leads to integration of the plasmid into the cellular chromosome.

7. precise excision, imprecise excision

8. cointegrates, double-strand breaks

9. Transposition of Mu requires that four subunits of the MuA transposase bind to and synapse the ends of the viral DNA. The synapsis is due to formation of a transposase tetramer with two subunits bound at each end of the viral DNA. This is followed by the nicking of each end of the DNA by a transposase subunit that is bound to the other end of the viral DNA. After nicking the target DNA, transposase subunits act again in trans to join the ends of the viral DNA to the ends of the target DNA.

10. acentric chromosomes, dicentric chromosomes

11. epigenetic effects

12. hybrid dysgenesis

13. TnpA recognizes the inverted repeats at the ends of the element and carries out the breakage and reunion steps of transposition in a way that generates a cointegrate. TnpR resolves the cointegrate by binding to the two copies of the resolution site and mediating strand breakage and exchange events within these sites.

14. phenotypic variegation

15. Pairing of the two RNAs blocks the binding of the ribosome to the IN RNA, thereby preventing translation.

B. Analytical and critical thinking

1. IS elements are composed of a transposase gene flanked on both ends by inverted repeats. Composite transposons are made up of two IS elements in either direct or inverted orientation relative to one another plus one or more genes that lie between the two IS elements. Transposase genes are found within the IS elements in composite transposons. The additional genes between the IS elements often encode antibiotic resistance traits or other traits with selective advantages for the cells. Transposons of the TnA family are flanked by short inverted repeats. Between these are the TnpA and TnpR genes required for transposition and resolution, the resolution site, and one or more genes for a selective trait.

2. DNA methylation changes the three-dimensional shape of DNA and thus the structure of the binding sites for proteins such as RNA polymerase and transposase. In this case the bacterial DNA is methylated at GATC sites within the Tn10 sequence by the cellular Dam methylase. Methylated DNA cannot be transposed. Immediately after replication the DNA is hemimethylated for a short period of time. During this time the transposase gene is transcribed and the mRNA translated. The unstable transposase then binds to the inverted repeats and can carry out the nicking and rejoining reactions required for transposition.

3. Nonautonomous elements are closely related in sequence to autonomous ones of the same type except that they cannot produce the proteins required for transposition and are therefore dependent on autonomous copies of the same element to provide these proteins. Structurally, nonautonomous elements must still contain the inverted repeat regions required for transposition.

4. The mRNAs encoding the repressor and the transposase proteins are the alternatively spliced products of the same primary transcript. Intron three is not removed in the repressor mRNA but is in the transposase mRNA. The repressor protein is smaller than the transposase because intron three contains an in-frame stop codon to terminate translation of the protein.

5. The transposase makes staggered strand breaks in the target site. The distance between the cleavage sites in the two strands will determine the length of the direct repeats that flank the two ends of the transposon. The ends of the single-stranded regions are joined to the ends of the transposons IRs. This leaves a gap at each end of the element that is filled in by DNA polymerase.

6. Target site selection is a function of the transposase, therefore the selection of preferred target sites depends on the binding of the transposase to these sites and not to random DNA sequences in the genome. Some elements insert randomly because the transposase will bind to any sequence with the same affinity. Other transposases will bind more strongly to certain sequences. This will lead to these sequences being the preferred target sites.

C. Extensions and applications

1. Conjugative transposons must have sequence features that define the ends of the transposon. These are not yet known for the CTnGERM1 transposon but are likely to be inverted repeats as are found on other DNA transposons. There must be a gene that encodes a transposase to nick the target site and to join it to the transposon DNA. Conjugation requires genes and sequences to mobilize the DNA and to transfer it to the recipient cell. Conjugative transposons like CTnGERM1 are often associated with drug resistance phenotypes that provide a selective

advantage to the cells that have the transposon when they are exposed to the appropriate drug. In addition to the *erm*G gene, the partial sequence data for this transposon includes several other putative genes that have sequence similarity to genes involved in other drug resistance phenotypes. To show an integration requirement the authors analyzed the DNA from a number of transconjugant strains that had independently acquired the element by conjugation. Twelve of twelve strains showed shifts in the size of restriction fragments by 75 kb, consistent with a 75 kb transposon integrating into the chromosomal DNA at various sites. None had an extra band of 75 kb that may have been present if the DNA was episomal. The very high sequence similarity to an erythromycin resistance gene previously known from a gram-positive bacterium indicates that the gene was likely to have been transferred recently to *Bacteroides* from a gram-positive bacterium. The G + C content of the transposon and of the *erm*G gene is more consistent with this direction of transfer rather than *Bacteroides* DNA being the origin of the *erm*G gene in *B. sphaericus*.

2. One mechanism may depend on the unstable nature of the transposase protein. Perhaps when bound to the transposon DNA the transposase has only a short time in which to act before it is disabled and can no longer carry out the activities required for transposition. Alternatively, the chromatin structure may influence the ability of the protein to function, remember that chromatin is looped and coiled and complexed with histone and nonhistone proteins. If one already has some positional data for a locus that is being targeted for mutation, then an approach that will work efficiently at only a short distance will generate fewer hits elsewhere in the genome. It should be easier to confirm that the phenotypes observed are due to the desired mutations rather than the results of mutations at second sites. A low transposition frequency would also decrease the likelihood of multiple transpositions in a single cell. Hence the phenotypes observed in mutants are likely to be the result of single transposition events. The mutants with partially active genes still had integration events but the integrations were near the 3 end of the gene so the protein products were almost the correct length and sequence and retained some function. Since there were short direct repeats derived from the original target site formed during the integration of the transposon, these made recombine to regenerate a "precise deletion" of the transposon. Alternatively since there is a homologous chromosome in the cell that has not suffered a transposition generated double-strand break there may be normal double-strand beak repair. Nonhomologous end joining is a third possible repair mechanism.

D. Standardized test format

1. C
2. B
3. D
4. E
5. D
6. A
7. C
8. B
9. E
10. A
11. C
12. E
13. A
14. D
15. B

Chapter 17

Retroviruses and retroposons

General Concepts	Section	Related Questions
RNA can be a transposition intermediate	17.1	B2, B4, D1
Retroviruses use transposition-like events	17.2, 17.5	A2, B4, D8
Retroviral RNA undergoes reverse transcription	17.4	A3, B4, D7
Retroviruses may transduce cellular genes	17.6	A5, D3
Viral superfamily retrotransposition	17.7-17.8	A6, B2, C1, D12
LINE retrotransposition	17.12	A13, A14, C2, D15
Processes/structures		
Genetic organization of retroviruses	17.2-17.3	A1, B3, B6, D2, D4, D6
Generating retroviral DNA forms	17.4-17.5	A3, A4, B6, D4, D5, D7
Retroposons resemble retrovirus DNA forms	17.7-17.8	A7, A11, B4, D9
SINE and LINE structures	17.9-17.12	A12-A15, D12
Processed pseudogene formation	17.11	A10, D14
Examples, Applications and Techniques		
HIV	17.3	B1, D2
Yeast *Ty* elements	17.7	A6-A8, B2, D9, D10
Drosophila copia elements	17.8	A8, B5, D11
SINES and LINES	17.9	A8, A11, A13, C2, D12
Alu family	17.10	A9, D12, D13
Human L1 LINE	17.12	A14, A15, D12, D15

Questions:

A. Recall and short answer

1. Although quite small, retroviral genomes encode numerous functional protein products. _____, _____ and _____ are used to produce alternative polypeptides from the viral primary transcript. These polypeptides are actually _____ that are cleaved into the final functional protein products.

2. The enzyme _____ is responsible for integration of retroviral DNA into cellular chromosomes. Its action is similar to that of the _____ that is required for the movement of transposons.

3. The order of the R, U5, and U3 sequences in a retroviral or retroposon LTR is _____.

4. Both minus and plus strand synthesis during retroviral reverse transcription require strand switching. This is an example of _____ mechanism of recombination.

5. Retroviruses that carry genes that can cause cancer in infected animals are _____. Often the genomes of these viruses lack the *pol* and *env* genes; as a consequence they are _____ and require a _____ to provide the missing viral functions.

6. Recombination between the two LTR's of a yeast *Ty* element leaves a 330 bp _____ as a molecular footprint in the genome.

7. What evidence suggests that the *Ty* elements in yeast have evolved from retroviral genomes?

8. What are two major differences between LINES and *Ty* or *copia* elements?

9. There are _____ copies of the AluI sequence in a haploid human genome. On average there is one AluI sequence in every _____ of human DNA.

10. Why is a processed pseudogene nonfunctional?

11. Retrotransposons of the _____ superfamily have LTR's at their termini but those of the _____ superfamily do not.

12. In the mouse genome SINES are typically derived from _____.

13. The proteins translated from LINES RNA are described as cis-acting, what does this mean?

14. Numerous partial LINES elements are present in eukaryotic genomes. Which region of the complete element is most commonly missing, why?

15. There appear to be _____ active LINES elements in the human genome. Are any other transposons or retrotransposons active in the human genome?

B. Analytical and critical thinking

1. What are the functions of the *gag*, *pol*, and *env* genes in a retroviral genome such as the HIV genome? Which of these genes are present in retroposons?

2. Describe the experiment that was used to show that the yeast *Ty* element used an RNA intermediate for its transposition.

3. In retroviruses and retroposons suppression of a stop codon at the end of the *gag* coding region or ribosomal frameshifting upstream of the *gag* stop codon are mechanisms used to regulate translation of the *pol* coding region. What does each of these mechanisms entail? What is the selective advantage for keeping *pol* expression low?

4. What features of retrotransposons resemble transposons? What features are different?

5. One feature observed with the *Drosophila melanogaster copia* element is a dramatic increase in transposition frequency when cells are grown in tissue culture. Mobile genetic elements in other species also may display this characteristic. Propose an explanation for the increased transposition frequency.

162

6. What are the functions of the LTR's in retrotransposons?

C. Extensions and applications

1. Transposable genetic elements account for about 45 to 50% of the human genome but only about 3% of the *Saccharomyces cerevisiae* genome and about 10% of the genome of the cellular slime mold *Dictyostelium discoideum*. A major difference between the human and the microbial genomes is in the compactness of the microbial genomes. There are many fewer and much smaller introns in the microbial eukaryotes than in man. A much greater proportion of the genome is composed of regions that encode proteins in the microbes. There is about one gene per 2.6 kb in *D. discoideum*. Winckler and coworkers (2002 *Cell. Mol. Life Sci.* 59:2097-2111) discuss the mobile genetic elements found in the cellular slime mold genome and compare them to those found in other eukaryotes. Of particular interest are the target site preferences of the various mobile genetic elements in cellular slime molds and other microbial species. Specific retrotransposons target the regions within 50 base pairs upstream from tRNA genes, the regions within 100 base pairs downstream of tRNA genes, or the less gene-rich regions near telomeres. For example, 75 % of the tRNA genes on chromosome 2 of *D. discoideum* have retrotransposons associated with them.

Discuss the adverse genetic effects expected from random transposition of mobile genetic elements. In which genome, human or slime mold, will these effects be greater, why? How does the development of a tRNA-targeted transposition mechanism mitigate some of the harmful effects of transposition? Propose mechanisms that may allow a retrotransposon to target the regions next to tRNA genes as integration sites.

2. Retrotransposons are important components of the human genome. The L1 (LINE-1) retrotransposon comprises about 17 % of our DNA. Recent estimates suggest that there are 60 to 100 elements capable of retrotransposition in the human genome. These encode an RNA binding protein and an endonuclease/reverse transcriptase protein. The proteins generally bind to the mRNA from which they have been translated and then catalyze integration of a new copy of the L1 element using their own mRNA as template for reverse transcription. However, the proteins can work in trans and are thought to mediate Alu transposition and processed pseudogene formation. There is a preferred target site with the consensus sequence 5′ TTTT/A 3′ for nicking by the endonuclease. L1 transcripts begin either at a specific site determined by an internal promoter within the element or at sites determined by promoters in the DNA upstream of the element. L1 transcripts end in 3′ Poly(A) tracts. The Poly(A) addition sites are either one encoded within the element or a downstream site in the DNA flanking the 3′ end of the element. Gilbert et al. (2002, Cell 110:315-325) and Symer et al. (2002 *Cell* 110:327-338) have developed methods to identify and characterize new L1 retrotransposition events in cultured human cells. They found many genomic deletions, the formation of chimeric elements, and comobilization of 5′ or 3′ sequences from the donor site to the new target site. These findings suggest that L1 may have a major role in shaping the evolution of the human genome.

Describe the target site primed reverse transcription process used for L1 retrotransposition. LINES are thought to be the only active mobile genetic elements in human genomes. How does the L1 retrotransposition mechanism help maintain functional elements? What processes are involved in comobilizing 5′ and 3′ sequences that flank an L1 element? What role may comobilization of these sequences play in the evolution of the human genome?

Formation of chimeric elements involves recombination between the newly integrating element and an element already located in the chromosome upstream of the target site of the new element. This process is accompanied by deletion of the intervening region of the chromosome. Propose a mechanism to account for the formation of chimeric elements. At new target sites there may be small deletions or small target site duplications. The deletions and the target site duplications are of variable lengths. Explain how the process of second strand nicking of the target site may generate either outcome.

D. Standardized test format

1. The organisms in which retroposons are found include

    A. eubacteria.
    B. gram-positive bacteria.
    C. archaebacteria.
    D. eukaryotes.
    E. gram-negative bacteria.

2. Which of the following has both an intracellular and an extracellular form?

    A. *copia*
    B. an HIV genome
    C. a *Ty* element
    D. a SINE
    E. a LINE

3. A viral gene with the ability to cause oncogenesis is

    A. the *pol* gene for reverse transcriptase.
    B. a proto-oncogene.
    C. the retroviral *gag* gene.
    D. a *c-onc* gene.
    E. a *v-onc* gene.

4. Retroviral genomes in cellular chromosomes are distinguished by the presence of

    A. inverted repeats at the ends of the genomes.
    B. microsatellites at the ends of the genomes.
    C. LTR direct repeats at the ends of the genomes.
    D. target site inverted repeats flanking the ends of the genomes.
    E. the delta sequence element.

5. The enzyme responsible for synthesis of the DNA copies of a retroviral genome is

    A. reverse transcriptase.
    B. DNA polymerase.
    C. RNA polymerase.
    D. transposase.
    E. integrase.

6. The RNA genomes in retroviral particles contain the

   A. LTR sequence at both the 5 and the 3 ends of the RNA.
   B. R-U5 segment at the 5 end and the U3-R segment at the 3 end of the RNA.
   C. LTR sequence at only the 5 end of the RNA.
   D. LTR sequence at only the 3 end of the RNA.
   E. long inverted repeats at each end of the RNA.

7. The initial primer for reverse transcription of retrovirus RNA is

   A. an RNA about 10 bases long made by the primase enzyme.
   B. the 3 end of the RNA which loops back to pair with a region of the RNA.
   C. the 3 OH group on a transfer RNA.
   D. a priming protein with an attached nucleotide.
   E. a nicked strand in cellular DNA which pairs with the viral RNA.

8. Which of the following is true?

   A. Retroviral integration into DNA typically does not generate target site repeats.
   B. Retroviral integration into DNA requires a circular viral genome.
   C. A complete linear retroviral genome is integrated into cellular DNA.
   D. Retroviral integration generates target site repeats and involves loss of some nucleotides from each LTR during the integration.
   E. Integration and reverse transcription of the viral RNA are directly coupled, since a nick in cellular DNA is the primer for reverse transcription.

9. The translation of the TyA-B protein in yeast requires

   A. an alternatively-spliced mRNA to be made.
   B. suppression of a stop codon to translate the TyB open reading frame.
   C. the use of alternative promoters to translate the TyA and the TyA-B mRNAs.
   D. the use of an antitermination signal to allow synthesis of the TyA-B mRNA.
   E. a ribosomal frameshift to translate the TyB open reading frame.

10. Messenger RNA produced from the *Ty* elements in a typical haploid yeast cell constitutes _____% of all the mRNA in the cell.

    A. greater than 5
    B. about 1
    C. 0.1
    D. 0.01
    E. 0.001

11. A major type of retrotransposon in *Drosophila* is the

    A. P elements.
    B. *copia* elements.
    C. *Ty* elements.
    D. LINES.
    E. IS elements.

165

12. Relic DNA transposable elements and retroposons make up about _____% of the human genome.

    A. 50
    B. 25
    C. 10
    D. 5
    E. 1

13. Human AluI sequences are derived from

    A. RNA polymerase I transcripts of the 7SL RNA.
    B. RNA polymerase II transcripts of the 7SL RNA.
    C. RNA polymerase III transcripts of the 7SL RNA.
    D. LINES by recombination between the terminal LTR's.
    E. the autonomous large universal sequence element.

14. Processed pseudogenes can be distinguished from pseudogenes that arise by mutational inactivation of genes by

    A. the presence of promoters, introns, and 3 polyA tracks in the processed pseudogenes but not in the mutated pseudogenes.
    B. the presence of promoters and introns in the processed pseudogenes but not in the mutated pseudogenes.
    C. the presence of introns and 3 polyA tracks in the processed pseudogenes but not in the mutated pseudogenes.
    D. the presence of promoters and introns in the mutated pseudogenes but not in the processed pseudogenes.
    E. the presence of 3 polyA tracks in the mutated pseudogenes but not in the processed pseudogenes.

15. Human L1 LINES integrate in cellular DNA via

    A. the action of a transposase encoded in the element.
    B. the action of an integrase encoded in the element.
    C. a site-specific recombination event carried out by an enzyme encoded in the element.
    D. unequal crossing-over reactions carried out by an enzyme encoded in the element.
    E. an endonuclease produced nick that serves to both prime reverse transcription and to join the element to cellular DNA.

Answers:

A. Recall and short answer

1. alternative splicing of the primary transcript RNA, frameshifting, or the suppression of a stop codon during translation; polyproteins

2. integrase, transposase

3. U3-R-U5

4. the copy choice

5. transducing viruses, replication-defective, helper virus

6. delta sequence

7. Critical evidence includes: 1) the common structural features shared by retroposons and retroviruses, 2) the common mechanisms involved in their action such as reverse transcription, gene expression and regulation, polyprotein formation, and integration, and 3) the presence of virus-like particles in yeast cells that contain Gag protein, *Ty* RNA genomes, reverse transcriptase, integrase, and double-stranded DNA copies of the genomes.

8. The two major differences are: 1) the absence of LTR's in LINES versus the presence of LTR's in the *Ty* and *copia* elements, and 2) the reverse transcription/integration of DNA copies is markedly different.

9. Early estimates put the number of AluI elements at 300,000 with one element per 6 kb; a more recent estimate based on the human genome sequence data puts the number of AluI elements at 1,300,000 with one element every 1.3 kb.

10. It cannot be expressed because it lacks a promoter. Therefore there is no transcript.

11. viral, nonviral

12. RNA polymerase III transcripts of the 7SL and tRNA genes

13. The translated proteins assemble on the messenger RNA that was used for their own translation rather than diffusing through the cell to complex with a different copy of the RNA. The protein-RNA complex is then translocated across the nuclear membrane where a copy of the LINES DNA is inserted into the genome.

14. The 5 end is usually missing because reverse transcription starts at the 3 end and proceeds until it either encounters the 5 end of a complete RNA or a 5 end generated by a strand break in the template RNA.

15. An early estimate of the number of active LINES elements was 10 to 50, data from the human genome sequencing projects indicates that 60 to 100 active elements are present. There does not appear to be any other active transposable element in the human genome.

B. Analytical and critical thinking

1. The Gag products bind the viral RNA and form a capsid structure. The Pol products are enzymes with reverse transcriptase, integrase, RNaseH, and protease activities. The Env products are membrane proteins required for budding of the virus and for infection of new cells. Retroposons have the *gag* and the *pol* genes but lack the *env* genes.

2. A *Ty* element was cloned and two changes made in its sequence. A mutation was placed in the U5 unique region that forms part of the upstream LTR and an intron was added into the sequence. The element was placed under the control of an inducible promoter, introduced into yeast cells, and transcribed. Transposition resulted in new copies of the introduced element that lacked the intron and also had the U5 mutation in both the upstream and downstream LTR. The only process known to precisely remove introns is RNA splicing. Therefore splicing must have occurred on the transcripts from the mutated element. The transfer of the mutation to the downstream LTR in the new copies of the element indicates that the DNA copies of the element are derived in a manner similar to that used by retroviruses.

3. Suppression of a stop codon involves the occasional binding of a normal tRNA to the stop codon in place of the binding of a translation release factor. The product generated in this case has the Pol peptide attached directly to the normal Gag peptide as a fusion protein, a polyprotein. For ribosomal frameshifting to occur the ribosome must shift position along the mRNA before the release factor binds and terminates translation. Typically the product from ribosomal frameshifting does not have the carboxyl end of the Gag protein present in the polyprotein. Instead there are other amino acids joining the Gag and Pol regions of the peptide due to the alternative frame that was used for their translation. The product is still a polyprotein. The products of the *pol* gene are all enzymes, while those from the *gag* gene are structural proteins. The virus needs a lot less of the enzymes. A low expression of the *pol* gene products preserves energy and amino acid precursors for other cellular or viral uses.

4. Retrotransposons are similar to transposons in having variable numbers of elements at variable locations within the genome of the host cell. Genetic effects are similar; they can cause insertional inactivation of genes and may be involved in DNA rearrangements based on recombination and other processes. Like transposons they also typically are surrounded by target site direct repeats. Even though the retroposons themselves end in direct repeats, the LTR's, the last few 5 and 3 bases in the LTR's form small inverted repeats. Differences include the genes encoded on the elements and the transposition mechanism.

5. Cells in culture, whether they are cells from animals or even microorganisms, are not in a natural environment. They are not receiving all of the normal signals from the environment or from other cells in the body. The cells typically are in a rich environment with normal selective pressures removed but others imposed. These changes will affect expression of the proteins required for transposition and lead to a change in transposition frequency as a consequence.

6. The LTR's are required for integration of the viral genome into DNA. The integrase protein must recognize and bind to the ends of the LTR's where it cleaves the DNA in preparation for joining it to the target site. The upstream LTR also provides the promoter required for transcription of the viral RNAs, the downstream LTR may activate transcription of cellular genes flanking the element.

C. Extensions and applications

1. Adverse genetic effects due to transposition include both insertional inactivation of genes and subsequent DNA rearrangements based on unequal crossover events that use the transposable elements as dispersed homologous sequences, for instance deletion of a segment of a chromosome following recombination between two copies of a mobile genetic element. Insertional gene inactivation is potentially a much greater problem in cellular slime mold genomes, with one gene every 2.6 kb and a higher proportion of protein-coding sequences, than in human DNA. There is a much higher proportion of non-coding DNA in the human genome that will not have its function disrupted by integration of a copy of a mobile genetic element. The regions just upstream and downstream of tRNA genes are intergenic regions, typically devoid of coding function. Therefore integration into these regions will not disrupt a gene. The tRNA genes also have internal promoters so that the gene will not be separated from its regulatory regions by integration events near to but outside of the tRNA coding region. Targeting to a noncoding region reduces insertional inactivation of genes but would not reduce effects due to recombination events between elements. One mechanism is for the integrase or endonuclease of the retrotransposon to target a sequence common to most or all tRNA genes, for example a sequence used as part of the internal tRNA promoter. A more plausible mechanism with some supporting evidence is based on protein-protein interactions of the retrotransposon proteins with transcription factors bound to the tRNA promoters. Different interacting proteins account for the upstream targeting by some elements and the downstream targeting by others.

2. The endonuclease binds to the new target site and cleaves one strand of the DNA 3′ to the track of T's in the 5′ TTTT/A 3′ site. The poly T track base pairs with the poly(A) track on the 3′ end of the L1 transcript and serves as the primer for the reverse transcription of the RNA. A nick is also made in the second strand of the DNA near to the target site. The 3′ end produced by this nick serves as the primer for synthesis of the second strand of the new L1 element. The cis-acting nature of the L1 proteins targets a functional mRNA to serve as the template for generating a new element in the genome. Thus there is a good chance that the new element also will be functional. Other elements that do not have mechanisms to pick out functional elements for copying are more prone to transpose nonfunctional elements. Eventually nonfunctional elements are the only ones remaining and further transposition events do not occur. The 5′ or 3′ comobilized sequences must be included in the reverse transcribed RNA. Therefore 5′ comobilization involves the use of a promoter upstream of the element and 3′ comobilization involves the use of a poly(A) addition site downstream of the element. The comobilized sequences are genomic DNA. They can be exons or promoters or any other sequence in the genome. These processes may be a key method for "exon shuffling" in eukaryotic genomes. Chimeric elements may arise by the interaction of the newly reverse transcribed strand with an element already in the genome. The reverse transcribed strand is a single strand DNA homologous to any other L1 element in the genome. Therefore the enzymes involved in homologous recombination can catalyze the formation of the chimeric elements, which may lead to the deletion of the DNA between the two sites on the chromosome. The nicking of the second strand does not occur at a specific site; it may be upstream or downstream of the nick in the first strand. It may also be at variable distances from the first strand cleavage site. Nicking to one side leads to deletions and nicking to the other side leads to target site duplications.

D. Standardized test format

1. D
2. B
3. E
4. C
5. A
6. B
7. C
8. D
9. E
10. A
11. B
12. A
13. C
14. D
15. E

Chapter 18

Rearrangement of DNA

General Concepts	Section	Related Questions
DNA rearrangements control gene expression	18.1-18.2	A3, B2
Switching changes the yeast mating type locus	18.2-18.9	B1, D1, D3, D5-D7
Rearrangement alters VSG expression	18.10-18.12	A6, B4, D2, D8
T-DNA transfer to plants causes disease	18.13-18.15	B3, D4, D9, D14
DNA amplification increases gene expression	18.16	A14, B5, C2, D10
Cells may be transfected with exogenous DNA	18.17-18.20	A10, B6, C1, D11-D13
Processes/structures		
Mating type switching	18.5, 18.8	A1, D1, D3, D7
VSG gene switching	18.11	A7, B4, D2, D8
Transfer of T-DNA	18.15	A8, D4, D9, D14
DNA amplification	18.16	A14, B5, C2, D10
Transfection	18.17, 18.18	A10-A13, C1, D11-D13
Examples, Applications and Techniques		
Yeast mating type genes	18.2-18.9	A2-A5, D3, D5, D15
Trypanosome VSG genes	18.10-18.12	A6, A7, B4, D2, D8
Ti plasmid	18.13-18.15	A8, A9, B3, D4, D9, D14
Dhfr gene amplification	18.16	B5
Transfection of embryonic stem cells	18.18-18.19	A10-A13, B6, D13
Gene replacement	18.20	A13, A15, C1

Questions:

A. Recall and short answer

1. Cells of yeast that can switch mating type carry a dominant allele for the _____, while those that cannot switch carry a recessive allele.

2. Mating in yeast requires cells of opposite mating types. How do the cells of the two types differ?

3. The proteins produced from the yeast mating type locus are _____.

4. In a colony produced from a single haploid yeast cell in which switching has taken place most of the cells are actually diploid cells produced by the fusion of cells of opposite mating type and then the subsequent cell growth and division of the diploid cells prior to meiosis and spore formation. What is the advantage to being diploid?

171

5. What proteins are required to maintain the silencing of the nonexpressed *HML* and *HMR* copies of the mating type loci?

6. The RNA polymerase that transcribes the active VSG gene is _____.

7. The inactive, nonexpressed forms of the variable surface glycoprotein genes in trypanosomes are called _____, while the active form of the gene is called the _____.

8. What are the functions of the VirA and VirG proteins in the two component system used by *Agrobacterium tumefaciens* in transferring the T-DNA from the Ti plasmid into plant cells?

9. The _____ are derivatives of the amino acid arginine that are produced by plant cells that carry T-DNA genes transferred to them from the Ti plasmid.

10. The introduction of exogenous DNA into recipient eukaryotic cells is termed _____.

11. In transient transfectants the introduced DNA exists in an _____ form; in stable transfectants it exists in an _____ form.

12. _____ animals have genetic material derived from another species.

13. What is the purpose of including genes such as that for G418 resistance on the transfecting DNA?

14. Circular extrachromosomal amplified DNA in mammalian cells take the form of _____, tandemly repeated gene amplifications are termed _____.

15. The genetic process that is involved in the replacement of an endogenous gene by a transfected gene is _____.

B. Analytical and critical thinking

1. Rearrangement of genomic DNA is used to control expression of the mating type gene in yeast cells. Based on what you have learned, what are the minimum genetic requirements needed to accomplish switching of the mating type genes in yeast cells?

2. Under what conditions is gene rearrangement used to control gene expression? In animals control of gene expression by gene rearrangement is rare, what is one example of gene rearrangement in the human genome? Is this rearrangement equivalent to what occurs during yeast mating type switching?

3. The disease caused by the Ti plasmid is crown gall disease. How are the genes in the T-DNA involved in gall formation? Why are the remaining genes on the Ti plasmid needed?

4. Why is the Variable Surface Glycoprotein (VSG protein) so important for trypanosomes? Why is switching of the VSG protein essential for a successful infection? What happens to trypanosomes that do not switch the VSG protein that is expressed on their cell surfaces or which switch to a VSG protein that is antigenically similar to that of a formerly-expressed VSG protein?

5. Cancers are often the result of overexpression of oncogenes while cancer treatment often involves the use of chemotherapeutic agents that target replicating cells. DNA amplification in human cells has major implications for our health, specifically in the occurrence and treatment of certain types of cancers. Describe how DNA amplification may lead to the production of cancer cells and how DNA amplification may make cancer cells refractory to chemotherapeutic treatment.

6. How are embryonic stem cells (ES cells) used in making transgenic animals? Why is using ES cells more efficient and more ethical than using egg cells to generate transgenic animals?

C. Extensions and applications

1. In *Drosophila melanogaster* the P element has been utilized to develop a sophisticated method to generate transgenic flies. Modified P elements that are composed of only the IRs needed for transposition plus the gene to be inserted into the genome are prepared. These are coinjected with wild type P element DNA into embryos. Transposase produced from the wild type P element mRNA can transpose either the wild type or the genetically-modified P element. Transgenes inserted in this way typically have normal expression patterns in fruit flies.

A similar approach for germ-line transposition transgenesis of vertebrates is being developed based on a DNA transposon, Sleeping Beauty, that was originally identified in Salmonid fish but which is active in many vertebrates including human cultured cells. Dupuy and coworkers (2002, PNAS 99:4495-4499) describe their use of Sleeping Beauty derivatives to insert transgenes into mouse single cell embryos. Modified elements that carried the gene for Green Fluorescent Protein between the Sleeping Beauty IRs were constructed and injected with Sleeping Beauty mRNA into the embryos. About 45 % of the embryos in this experiment were transgenic and of these embryos one third had transposase-mediated integration events. In a second experiment with a different transgene 45 % of the embryos were again transgenic and of these two thirds had transposase-mediated events. Some animals had multiple transposase-mediated events and/or a mix of random insertions and transposase-mediated events. Unlike in flies, expression of the transgene was variable, sometimes it was not expressed or it was expressed in low amounts. Some of the sequence data obtained by Dupuy for the junction regions containing the ends of the Sleeping Beauty IRs and the flanking insertion sites is shown below. The IR sequence is shown in bold face type.

ACTTCAACTGTATAGAAACA   **ACTTCAACTG**TAGGTCCTAG
ACTTCAACTGTATGTGTGGG   **ACTTCAACTG**TAGATACTCA
ACTTCAACTGTATATCCAGG   **ACTTCAACTG**TAACTCAGCT
ACTTCAACTGTAGCTACAAA   **ACTTCAACTG**TAAGTCTCCC
ACTTCAACTGTATATAGGCC   **ACTTCAACTG**TAAAATGCAT

Is there a target site preference? Since the different insertions of the transgene all have the same promoter and are not altered in sequence, why does the expression of the transgene vary? What is one advantage of injecting mRNA rather than copies of the wild types transposon? When the Sleeping Beauty element transposes the ends of the old target site are efficiently repaired to produce sites that have one of two possible sequences TACAGTA or TACTGTA. If the two TA dinucleotides at the ends of these sequences are derived from a target site direct repeat, propose a mechanism to account for the generation of these excision sites.

2. Recent work has investigated some of the initial steps in the development of gene amplifications. Shimizu and coworkers (2003 *Cancer Research* 63:5281-5290) studied the role of replication initiation regions and of 'controllable conflicts between replication and transcription' in generating amplifications in tissue cultured cells. Their work began with replication initiation regions (probable replication origins) from previously known amplified regions cloned in conjunction with drug resistance traits on recombinant DNA plasmids. They observed the generation of large plasmid multimers composed of tandem arrays of the initial plasmid sequences. These were similar in size to native double minute chromosomes. They found that these integrated into chromosomes and that the chromosomes then went through a breakage-fusion-bridge cycle similar to that observed with the chromosomes broken by movement of Ac elements in maize. A key point of their findings was the importance of strand breaks for the integration of the plasmid DNA into a chromosomal site. These were formed by the conflict between replication and transcription both of which are generating single-strand template DNA that may be more easily broken to generate single-strand regions with 3′ ends such as are used by RecA protein in *E. coli* as a substrate for recombination.

Propose a model for amplification of a segment of DNA based on an initial unequal crossover event that involves copies of transposable genetic elements to generate an extrachromosomal circle and subsequent events to generate a double minute chromosome and an HSR region. Why is the replication initiation region required? Why are single-stranded regions with 3′ ends needed? What will be the structure of an HSR region generated via repeated breakage-fusion-bridge cycles? Why are HSR amplifications maintained more stably in the absence of selective pressure than amplifications based on double minute chromosomes?

D. Standardized test format

1. When mating type switching takes place, what happens to the DNA that originally occupied the mating type locus?

 A. It is degraded into nucleotides.
 B. Most of it remains intact only a few nucleotides are switched by a mismatch repair system.
 C. It is moved into one of the silent loci.
 D. It is transferred to a cell of the opposite mating type by a conjugation-like process.
 E. It is transferred to the telomeric region of a chromosome where it no longer can be expressed.

2. The typical expression-linked site for the gene encoding the VSG protein of trypanosomes is located within

 A. a unique sequence region near the telomere of a chromosome.
 B. a repetitive DNA region near the centromere of a chromosome.
 C. a repetitive DNA region near the telomere of a chromosome.
 D. a unique sequence region near the centromere of a chromosome.
 E. a large extrachromosomal DNA with an inverted repeat structure.

3. Genes in which mutations eliminate the ability of yeast cells to mate are denoted *STE* (for sterile). Mutations in *STE2* and *STE3* are specific for the individual mating types but mutations in the other *STE* loci affect both mating types. The *STE2* and *STE3* loci encode

    A. the silent copies of the yeast mating type gene.
    B. transcription factors that are specific to each mating type.
    C. proteins that determine which silent locus will be used to replace the gene in the *MAT* locus.
    D. the cell surface receptors for the mating pheromones.
    E. proteins that specifically silence the *HML* and *HMR* loci.

4. The transfer of the T-DNA from the Ti plasmid to a plant cell involves identification of

    A. the T-DNA based on inverted repeats at each end of the T-DNA, nicking of one strand, transfer of a region from the nicked strand to the plant, conversion to a double-strand duplex, and integration into the plant genome.
    B. the T-DNA based on direct repeats at each end of the T-DNA, nicking of one strand, transfer of a region from the nicked strand to the plant, conversion to a double-strand duplex, and integration into the plant genome.
    C. the T-DNA based on inverted repeats at each end of the T-DNA, double-strand cleavage outside the inverted repeats, transfer of the double-strand DNA to the plant, and integration into the plant genome.
    D. the T-DNA based on direct repeats at each end of the T-DNA, nicking of one strand, transfer of a region from the nicked strand to the plant, and integration of the single-strand DNA into the plant genome.
    E. the origin of transfer (oriT), nicking of one strand, transfer of the entire Ti plasmid to the plant cell, conversion to a double-strand duplex, and integration of the circular Ti plasmid DNA into a unique site in the plant chromosome.

5. Combinations of the transcription factor PRTF, a1, α1 and α2 specifically activate or repress different genes. What types of genes are repressed by a combination of the a1 and the α2 proteins?

    A. diploid-specific genes
    B. mating type **a**-specific genes
    C. haploid-specific genes
    D. mating type α-specific genes
    E. the *SIR* genes

6. For proper function of the *HML* and the *HMR* loci the Origin Recognition Complex (ORC) must bind to an *ARS* site near the silent mating type loci. Why is ORC needed?

    A. Continued replication across the silent loci prevents their transcription.
    B. One or more of the ORC proteins aids the binding of the SIR1 protein that initiates silencing.
    C. Licensing factors that bind to ORC mediate formation of facultative heterochromatin at the silent loci.
    D. Replication beginning at the ORC is required to switch the mating type.
    E. The ORC is so big it covers much of the DNA in the silent loci, this prevents transcription of the silent loci.

7. The yeast HO endonuclease is expressed only in

    A. haploid cells during the G2 phase of the cell cycle.
    B. diploid cells just before the first meiotic prophase.
    C. haploid cells during the M phase of the cell cycle.
    D. diploid cells entering meiosis just before DNA replication begins.
    E. haploid cells at the end of the G1 phase of the cell cycle.

8. A trypanosome has _____ genes encoding different forms of the variable surface glycoprotein.

    A. about 1000
    B. about 100
    C. 3 (one active and two silent copies)
    D. about 20
    E. about 20,000

9. For transfer of T-DNA from the *A. tumefaciens* Ti plasmid to plant cells, genes and sequences from the Ti plasmid as well as the bacterial genes *chv*A, *chv*B and *psc*A are required. The bacterial genes encode

    A. an endonuclease and two exonucleases.
    B. a helicase, a polymerase, and a ligase.
    C. a response regulator system.
    D. proteins to manufacture a cell surface exopolysaccharide.
    E. proteins that confer resistance to phenolic plant defense compounds.

10. The *cob*A type of cobalt ion resistance in cellular slime molds is a stable recessive trait while the *cob*B type is an unstable dominant trait. Based on the typical features associated with chromosomal genes and with genes present on amplified DNA in eukaryotic cells which of the following is true?

    A. Both traits are due to extrachromosomal gene amplifications.
    B. Both traits are due to tandem arrays of amplified genes within chromosomes.
    C. the *cob*A trait is due to a mutation in a single copy chromosomal gene while the *cob*B trait arises from an extrachromosomal gene amplification.
    D. the *cob*B trait is due to a mutation in a single copy chromosomal gene while the *cob*A trait arises from an extrachromosomal gene amplification.
    E. Both resistance traits are due to mutations in single copy chromosomal genes.

11. Expression of transgenes in transfected animal cells is

    A. typically independent of the cell type or developmental stage.
    B. highly uniform from animal to animal despite differences in the integration site of the transgene.
    C. completely independent of the normal regulatory control mechanisms.
    D. often dependent only on the number of transgene copies in the cell.
    E. variable from animal to animal and dependent on the integration site.

12. The typical form of DNA in stably-transfected animal cells is

    A. a tandem array of many gene copies integrated into a chromosome.
    B. a large extrachromosomal multimeric DNA molecule.
    C. a large number of small circular extrachromosomal DNAs.
    D. an extra unique single copy gene in a chromosome.
    E. an extra unique single copy gene on an extrachromosomal DNA.

13. Transgenic mice that have been derived by fusion of transfected embryonic stem cells and normal blastocysts are

    A. sterile tetraploid animals.
    B. sterile diploid animals.
    C. heterozygous for the transgene.
    D. chimeric mixtures of the two cell genotypes.
    E. homozygous for the transgene.

14. The function of the Ti plasmid *vir*D1 and *vir*D2 genes is to encode

    A. an exonuclease that degrades plant DNA.
    B. an endonuclease that initiates T-DNA transfer.
    C. a two component response regulator.
    D. proteins required for opine synthesis.
    E. proteins required for opine degradation.

15. The *STE* (for sterile) genes that are involved in the mating of yeast cells encode

    A. transcription factors.
    B. proteins required to make the a and the α pheromones.
    C. proteins required to degrade the a and the α pheromones.
    D. proteins that mediate the fusion of the a and the α haploid cells during mating.
    E. proteins involved in the signal transduction pathway that responds to the presence of the mating pheromones.

Answers:

A. Recall and short answer

1. HO endonuclease, this endonuclease is required to recognize and produce a double strand break at the mating type locus.

2. Cells of mating type **a** carry the **a** allele in the mating type locus, produce the **a** pheromone, and have a cell surface receptor for the α factor. Cells of mating type α carry the α allele in the mating type locus, produce the α pheromone, and have a cell surface receptor for the **a** factor.

3. regulatory proteins that control the expression of genes

4. There are two copies of the genome and therefore there is an increased likelihood for the repair of damaged DNA. In colonies formed by mixing of the progeny from several different haploid cells there would also be the possibility of complementation of genetic defects in the diploids.

5. SIR1-4, RAP1, histone H4, and the ORC complex

6. RNA polymerase I

7. basic copy genes, expression-linked copy

8. VirA is a cell surface protein that becomes autophosphorylated when it binds a phenolic compound produced by a wounded plant. It in turn phosphorylates and activates the VirG protein. This allows VirG to bind DNA and activate transcription of the other *vir* genes on the Ti plasmid.

9. opines

10. transfection

11. extrachromosomal, integrated

12. transgenic

13. Only a small proportion of the cells exposed to the transfecting DNA will actually take in DNA. The selective agent is used to allow the preferential growth of cells that have acquired and can express the genes on the transfectingDNA by inhibiting the growth of or killing the nontransfected cells.

14. double minute chromosomes, homogeneously staining regions (HSR)

15. homologous recombination

B. Analytical and critical thinking

1. Three things are required. There must be a way to recognize the expressed copy of the mating type gene; this is the 24 bp sequence bound by the HO endonuclease. There must be nonexpressed copies to serve as the source of replacements for the expressed copy; these are the silent copies of the **a** and the α mating type cassettes. Finally there must be a mechanism to remove the expressed copy and replace it with a silent copy; this is accomplished by double strand breakage catalyzed by the HO endonuclease, followed by exonuclease removal of the previously expressed copy of the mating type DNA, and its replacement via double-strand break repair using one of the silent copies as the template for the DNA inserted into the expressed mating type locus.

2. Rearrangement may be used to create new genes or to switch expression from one pre-existing gene to another. The rearrangement of the immunoglobulin genes and the T-cell receptor genes are examples that occur in humans. These rearrangements are not equivalent to those that occur in yeast or in trypanosomes since the rearrangements occur in human somatic cells and are not reversible.

3. The genes in the T-DNA encode proteins that induce plant cell division and hence gall formation. They also encode proteins that synthesize plant hormones that induce differentiation of the cells in the gall into root and shoot cells. The T-DNA also contains genes for the conversion of arginine into one of the opines. Opines can be used as carbon and nitrogen sources by *A. tumefaciens* cells that carry the Ti plasmid. The other genes on the Ti plasmids include the genes for proteins needed to metabolize the opines, to recognize the presence of plant wounds, to regulate expression of the genes on the Ti plasmid, to transfer the T-DNA into the plant cells at the wound site, and to control the replication and copy number of the plasmid in bacterial cells.

4. The Variable Surface Glycoprotein (VSG protein) serves as a bulwark against the defenses of the host. It coats the cell surface and prevents access by the host immune system to other trypanosome antigens. The host produces antibodies against antigens in the VSG protein. Switching of the VSG protein type allows the trypanosome to evade the immune response mounted against earlier forms of the VSG protein since the immune response will be ineffective against the new VSG form. If the trypanosome does not switch to a new immunologically distinct form of VSG protein, then it will be killed by the immune system of the host.

5. Amplification of the native proto-oncogene generates large numbers of the gene and leads to overexpression of the gene product. The overexpressed protein is identical to the normal protein but present in much larger amounts. This leads to increased cell division. Chemotherapeutic agents typically inhibit normal cellular enzymes involved in replication or the production of nucleotides for replication since this is one feature that cancer cells do not share with most somatic cells. Drug resistance arises if the target enzyme of the drug is present in greater abundance than is normal. Hence amplification of the gene that encodes the target enzyme of the chemotherapeutic agent will allow the cancer cells to become resistant to the drug and refractory to the therapy. An example is the methotrexate resistance of cells that arises from amplification of the *dhfr* genes.

6. ES cells can be grown in tissue culture indefinitely and then transfected with a genetic construct. ES cells that have incorporated the transgene can be selected based on cotransfection with a gene encoding a selective trait such as G418 resistance. Alternatively, each transfected ES cell can be grown separately from the others, DNA can be isolated from a portion of each of the

resulting clones, and the DNA samples screened using PCR techniques. Thus ES cells with successful integration events can be identified and only these used in generating transgenic animals. With transfection of egg cells many animals will need to be screened before one with the required genotype is identified. Thus the cost of the experiment both monetarily and ethically is greater.

C. Extensions and applications

1. There is a preference for a TA target site. These are the TA dinucleotides that flank each of the IR sequences. Sleeping Beauty elements target the TA sequence and use it to generate a two base pair direct repeat that flanks each copy of the element. Expression appears to depend on the site of integration of the transgene, some sites are in regions that do not allow expression of the transgene due to the structure of the chromatin. If mRNA is injected rather than copies of the wild type transposon DNA, then there will be transposition only immediately after the construct is injected into the embryo. There cannot be additional rounds of transposition that may inactivate genes or cause other problems. The transposase must be generating staggered cuts in the IRs when it excises the transposon. This leaves a three base single-stranded region on each side of site. From the sequence of the last three bases in the IR you can predict that these single-stranded regions may form two GC base pairs with a mismatched pair between them. Sometimes the mismatch is repaired to an AT base pair and sometimes it is repaired to a TA base pair.

2. Copies of transposable genetic elements dispersed throughout chromosomes provide homologous sequences at which recombination may take place. If a recombination event takes place between elements located at two different sites on a chromosome, then the DNA between the transposable elements is deleted from the chromosome and generates a monomeric extrachromosomal circular DNA. If this carries a replication initiation region then it may be replicated during S phase and maintained in the cell. The daughter circles can recombine with each other to generate dimers. As subsequent rounds of replication and recombination take place larger and larger circles with more copies of the sequence will be generated, these are double minute chromosomes and will have the copies of the DNA segment arranged in a tandem array. Recombination between the circular molecules or between a circular molecule and an homologous chromosomal region is facilitated by the occurrence of strand breaks that may be generated when a replication fork and a transcription unit that are both unwinding the template DNA butt up against one another. Integration generates an HSR-like region with a tandem array of the amplified region. Many HSR's actually have a complicated arrangement of head to tail and head to head copies of the amplified DNA. Breakage of the chromosome in the HSR region generates chromosome fragments that may be joined by the nonhomologous end joining mechanism. Further rounds of breakage and rejoining generate more and more complicated structures within the HSR and may lead to an increase in the size of the HSR. The replication initiation region is required to allow the DNA segment to be replicated, if it cannot replicate, then it cannot increase in number. Single-strand regions are needed to allow complementary base pair formation between recombining DNA molecules. The HSR's generated by a breakage-fusion-bridge cycle will have mixtures of head to head and head to tail copies of the DNA segment. HSR's are more stably maintained than double minute chromosomes since they are physically linked to centromeres and will be segregated to both daughter cells at cell division. DNA molecules without centromeres are often unequally segregated at mitosis and hence lost from some daughter cells.

D. Standardized test format

1. A
2. C
3. D
4. B
5. C
6. B
7. E
8. A
9. D
10. C
11. D
12. A
13. D
14. B
15. E

Chapter 19

Chromosomes

General Concepts	Section	Related Questions
Genomes are packaged structures	19.1-19.7	
Gene expression is associated with a reduced packaged state	19.9-19.11	
The chromosome is a segregation device	19.12	
Specific sequences are required for chromosome segregation	19.13-19.15	
The ends of chromosomes need to be stabilized	19.16-19.19	

Processes/structures		
Viral genomes are packaged	19.2	D1, D2
Bacteria contain nucleoids	19.3	B4, C3, D2
Chromatin at interphase of eukaryotes	19.5-19.7	A1, A2, C2, C3, D2, D3, D4, D5
Mitotic and meiotic chromosomes	19.8	A3, A4, A5, A6, A8, D2
Gene expression occurs in extended DNA	19.9-19.11	A7, A9, A10, D6, D7
Chromosome segregation/centromere	19.12-19.15	B1, B2, C1, D8, D9
Chromosome ends-telomere	19.16-19.19	B3, C1, D10

Questions:

A. Recall and short answer

1. Define chromatin.

2. DNA in an interphase nucleus is observed as densely packaged chromatin called _____ and less densely packaged material called _____.

3. Giemsa is used to stain A. mitotic, B. interphase chromosomes.

4. What component of mitotic chromosomes is stained by the dye Giemsa?

5. Genes are more abundant in the A. Giemsa bands, B. interbands of mitotic chromosomes.

6. The GC content of DNA is A. higher, B. lower in the Giemsa bands of mitotic chromosomes.

7. What are lampbrush chromosomes?

8. Of what use is Giemsa staining?

9. What are polytene chromosomes?

10. What are Balbiani rings?

B. Analytical and critical thinking

1. Why are yeast plasmids unstable unless they include a centromere?

2. Describe the arrangement of components at the centromere. Include in your answer the centromeric DNA, microtubules, centromeric binding proteins, and microtubule binding proteins.

3. What effect would a mutation in telomerase have on chromosomes?

4. In the *E. coli* nucleoid the chromosome associates with protein H1. Propose a hypothesis to explain the observation that the H1 gene has been identified as a misregulation mutant of a few different genes.

C. Extensions and applications

1. Describe all the components required for a DNA cloning vector that can replicate very large DNA fragments in a eukaryotic cell.

2. There are two models for the arrangement of chromosomes in the interphase nucleus. One is the "spaghetti model," which depicts the chromosomes as being arranged in a mass like a plate of spaghetti. The other is the "Tokyo subway model," which depicts the chromosomes as being maintained in an orderly arrangement like a major metropolitan subway system. What evidence can you glean from Chapter 19 that supports the Tokyo subway model?

3. Explain how ethidium bromide can be used to measure DNA supercoiling.

D. Standardized test format

1. Which of the following functions is not required for genome packaging of lambda?

    A. translocation
    B. condensation
    C. DNA cleavage
    D. capsid binding
    E. recombination

2. How is the DNA packaging ratio calculated?

    A. the length of DNA ÷ length of the packaged unit
    B. length of the packaged unit ÷ the length of DNA
    C. the diameter of DNA ÷ by the length of DNA
    D. the weight of DNA ÷ weight of the packaged unit
    E. the length of DNA without associated proteins ÷ by the length of DNA with associated proteins

3. What is a "matrix attachment region"?

    A. region of chromosomes where genes are found
    B. region of chromosomes that attach to the microtubule matrix
    C. region of chromosomes needed for proper segregation during karyokinesis
    D. region of chromosomes that attaches to the nuclear matrix
    E. region of chromosomes where attachment to the nuclear membrane occurs

4. What is a chromocenter?

    A. regions where chromatin is located, for example, the nucleus
    B. chromatin regions that are always heterochromatic
    C. regions where chromatin changes from heterochromatic to euchromatic state
    D. regions where chromatin is attached to the nuclear matrix
    E. regions where heterochromatin is located

5. Which of the following statements about constitutive heterochromatin is false?

    A. It is permanently condensed.
    B. consists of small, repeated DNA sequences
    C. contains few genes
    D. has reduced frequency of recombinantion
    E. is very transcriptionally active

6. Which of the following is a false statement about how amphibian lampbrush chromosomes differ from meiotic chromosomes of other species?

    A. Lampbrush chromosomes are transcriptionally active, whereas most meiotic chromosomes are not.
    B. Lampbrush chromosomes have a much lower packaging ratio than meiotic chromosomes from other species.
    C. Lampbrush chromosomes consist of two pairs of sister chromatids, whereas meiotic chromosomes from other species consist of one pair of sister chromatids.
    D. Lampbrush chromosomes exist because the amphibians in which they are found can have meiotic periods lasting several months, whereas meiosis in other species is comparatively short.
    E. Lampbrush chromosomes exist for only a portion of the meiotic period, and at other times they are similar to meiotic chromosomes from other species.

7. Which of the following is a true statement comparing lampbrush chromosomes and polytene chromosomes?

    A. Both are meiotic chromosomes.
    B. Both are found in amphibians.
    C. Both have extended regions where transcription occurs.
    D. Both show banding patterns that can be associated with expression of specific genes.
    E. Both can be visualized with a light microscope.

8. Which of the following is a true statement about centromeres?

    A. Centromeres have heterochromatin that is rich in satellite DNA.
    B. Centromeres are always located in the middle of a chromosome.
    C. Only linear chromosomes require a centromere.
    D. Centromeres prevent chromosomal breaks and rearrangements.
    E. Centromeres link chromosomes to the nuclear matrix.

9. Which of the following is a false statement about the CEN element of *Saccharomyces cerevisiae* chromosomes?

    A. CDE-I and CDE-III are A-T-rich sequences that flank a short conserved sequence called CDE-II.
    B. CBF1 protein binds to CDE-1, but this interaction is not essential for centromeric activity.
    C. Cse4p bends the CDE-II when it binds and this resembles the structure of the nucleosome.
    D. The CBF3 protein complex that binds to CDE-III is essential for centromeric function.
    E. A group of proteins called Ctf19, Mcm21, and Okp1 link the proteins bound to CDE-I, CDE-II, and CDE-III.

10. Which of the following is a false statement about telomerase?

    A. Telomerase is a ribonucleoprotein and the RNA component base pairs with the telomere.
    B. The RNA component of telomerase serves as the template for telomere synthesis.
    C. Telomerase uses the 5'-end of the telomere to prime telomere synthesis.
    D. The protein component of telomerase is a reverse transcriptase.
    E. dCTP and dATP are not substrates of telomerase.

Answers:

A. Recall and short answer

1. Chromatin is the state of nuclear DNA with associated proteins in an interphase nucleus.

2. heterochromatin, euchromatin

3. mitotic

4. this is not known

5. interbands

6. lower

7. Lampbrush chromosomes are the extremely extended meiotic chromosomes in the oocytes of some amphibians.

8. Giemsa staining allows each chromosome to be identified by its characteristic banding pattern.

9. Found in some fly tissues, polytene chromosomes are large in that they are the result of many successive rounds of replication without separation of the products.

10. Balbiani rings are extremely large puffed regions on polytene chromosomes.

B. Analytical and critical thinking

1. Yeast plasmids without a centromere segregate erratically during mitosis and sometimes fail to move into a daughter cell during cell division.

2. The centromeric DNA is bound by the centromeric binding proteins, that are in turn linked to microtubule binding proteins, that link the entire structure to the microtubules.

3. First, the telomeres will become progressively shorter after each round of replication. Chromosomes will then break and rearrange.

4. Since H1 is involved in packaging the chromosome into the nucleoid it is possible that regulation of specific genes depends on the packaged state of the chromosome. For example, transcriptional regulators and RNA polymerase must have access to the operator and promoter sequences in order to regulate expression. Limiting access by associating the promoter and operator with packaging proteins would prevent the regulators from functioning. Also, packaging with H1 could alter DNA supercoiling in certain regions and supercoiling can affect the transcriptional activity of promoters.

C. Extensions and applications

1. The engineered chromosome would require an origin of replication that would function as the *cis* component of the host DNA replication machinery, a centromere to ensure that the artificial chromosome will segregate, and telomeres to prevent chromosome rearrangements and breaks.

2. The act of mitosis requires that very large segments of DNA must accurately disentangle from their state in the interphase nucleus and condense into mitotic chromosomes. Afterwards the mitotic chromosomes disperse into their state in the interphase nucleus. If the process were not orderly, active mechanisms must exist to disentangle the chromosomes. A more plausible explanation is that the arrangement of chromosomes is orderly.

3. Ethidium bromide is a fluorescent dye that intercalates between base pairs of DNA. In a closed circular DNA molecule the intercalation introduces positive supercoils. The amount of ethidium bromide required to achieve zero supercoiling is a measure of the original density of negative supercoiling. In a linear molecule, intercalation does not introduce supercoils because the DNA can rotate freely.

D. Standardized test format

1. E
2. A
3. D
4. B
5. E
6. C
7. C
8. A
9. A
10. C

> ## Chapter 20

Nucleosomes

General Concepts	Section	Related Questions
Histone core structure	20.1 - 20.4; 20.8	A1, A2, A3, A4, C1
DNA nucleosome association	20.5 - 20.7	A5, A7, A8, B3, B5, D2
Core and Linker DNA	20.4	A6, B4, D4
Locus control regions	20.16	A15, D1, D8
Chromatin	20.7	A7, B3, B5, C1
Processes		
Histone modification	20.9	A8, A9, C-2, D3, D6, D10
Chromatin reproduction / assembly	20.10	A10, B1, D9
Nucleosome positioning	20.11	A11, A12, B4, D7
Transcription / histone assembly	20.12-20.14	A13, B2, C1, C2, D5
Replication / histone assembly	20.10	A10, B1, D9
Examples, Applications and Techniques		
Hypersensitive site analysis	20.14	A14, B2, D5
Micrococcal nuclease treatment	20.4	A4,
DNAase I / II treatment	20.5, 20.15	A5, A15

Questions:

A. Recall and short answer

1. What are 3 levels of chromatin organization? How does packing ratio reflect the physical organization of the chromosome?

2. Physically, chromatin structure differs in organization throughout the cell cycle. What are the packing ratios of heterochromatin and euchromatin during interphase of the cell cycle? How does the structural organization reflect the ability of the DNA to be expressed? Does the packing ratio of specific chromatin segments within a cell change as a consequence of cellular differentiation? Explain.

3. What is the structure of the nucleosome core particle? How does the model for core particle association reflect the molar ratios of histone monomers present?

4. Treating chromatin with micrococcal nuclease could be expected to yield a 200 base pair ladder when examined using agarose electrophoresis. Under what DNA packing conditions would the possibility of this result exist?

5. Using DNAase I to nick free and nucleosome bound DNA, would one expect to find a difference in the resulting DNA pattern observed using agarose electrophoresis? How does the resulting pattern of DNA fragments reflect the organization of DNA?

6. _____DNA is of invariant length (146 bp) and is resistant to digestion by nucleases, while this DNA maintains a variable length and comprises the remainder of the repeating nucleosome unit_____ _____.

7. Chromatin is organized in the form of fibers. What structural configuration do these fibers assume?

8. The crystal structure of the histone core octamer suggests a model for core assembly. Is it probable that the core histone monomers interact with molecules outside the core octamer? Describe how the core assembly interacts with the DNA molecule.

9. The transient covalent modification of the N-terminal tails of the core histone proteins influences nucleosome structure. How does the covalent modification influence net charge of the core histone particle?

10. Normal histones are used to construct nucleosomes during _____, while variant histone proteins are required for _____nucleosome assembly. Why can't normal histones be used within all cellular processes?

11. The placement of a nucleosome in relation to DNA sequence is clearly not a random process. One view describing this process includes the _____model, which requires the sequence specific placement of each nucleosome. The second view is the _____. What are the features of this second model?

12. _____describes the location of DNA on the nucleosome with regard to the boundaries of the nucleosome, while _____describes DNA's position relative to the octamer surface. Under what circumstance is each relationship important in facilitating access to DNA?

13. Is it possible for transcribed genes to be organized into nucleosomes? Briefly describe.

14. Hypersensitive sites reflect what? How does the existence of a hypersensitive site correlate with the activity of the promoter region that contains it? How are nucleosomes excluded from these segments of DNA?

15. Susceptibility of DNA to DNAase I is a characteristic of all genes that are able to be transcribed. How does the 5' locus control region within domains reflect this relationship?

B. Analytical and critical thinking

1. Because you are interested in determining how nucleosomes assemble during replication, and you wish to determine whether the protein CAF-1 is involved within the process, an *in vitro* replication experiment is designed with and without CAF-1 present. Predict the results of the experiment in the presence and absence of CAF-1. How would the results of the experiment change if you were testing replication independent nucleosome assembly? Why?

2. (a) Consider the structure and location of hypersensitive sites within and around nucleosomes. Does the term hypersensitive site imply naked DNA? Explain. Describe the association between hypersensitive sites within the genome and selective gene expression.

(b) How does the DNA contained within hypersensitive sites rebuff nucleosome association?

(c) Considering the association between hypersensitive sites and gene expression, predict which classes of proteins would likely create/maintain hypersensitive sites.

3. What relationship exists between nucleosome monomer DNA content and histone H1 association? How does the relationship reflect the spatial association between histone H1 and the nucleosome core? Considering that histone H1 depleted chromatin is more readily solubilized, what function does this suggest for Histone H1 within chromatin?

4. (a) Is the association of DNA sequence with histone core octamers random? Describe the characteristics typical of core and linker DNA.

(b) Correlate a change in DNA position relative to the nucleosome core with the introduction of boundaries within the promoter region.

(c) Provide a model in which the inhibition of boundary development inhibits promoter activity.

5. Consider the structure of the chromatin fiber and the ability of RNA polymerase II to carry out transcription *in vitro*. When chromatin is placed into an environment with low ionic potential the nucleosomes present a strong barrier to RNA polymerase II; however, by increasing ionic strength of the solution a vast majority of the same DNA sequences were transcribed to completion. Explain these observations.

C. Extensions and applications

1. The unifying model describing displacement of histone core octamers during transcription by RNA polymerase II lacks a clear understanding of nucleosome disassembly and reassembly. Work by Belotserkovskaya et al. (2003 *Science* 301:1090-1093) described a dimeric protein complex (FACT - facilitates chromatin transcription) containing highly acidic domains, typical of histone chaperone proteins. Functional analysis of the protein complex *in vitro* demonstrated the ability to release H2A-H2B histone dimers and facilitate the complete transcription of DNA templates in solutions of low ionic strength. In addition, the protein complex demonstrated the ability to generate high molecular weight chromatin from core histone monomers and DNA.

(a) What biological function do chaperone proteins maintain?

(b) What effect does the chaperone protein complex FACT have on transcription?

(c) Considering the results, provide a model for FACT action during transcription.

2. The N-terminal tails of histones are modified by methylation, acetylation and phosphorylation resulting in the elimination of positive charges and the introduction of negative charges to the nucleosome core. The pattern of covalent modification has been termed the "Histone code" as the pattern of covalent modification contains valuable information regarding gene expression. Work by Jacobs and Khorasanizadeh (2002 *Science* 295:2080-2083) indicates that the methylation of the lysine at position 9 of histone H3 is required and sufficient to bind the HP-1 family of proteins, which lead to heterochromatin formation.

(a) What functional information is contained within the "Histone code"?

(b) Considering the influence of nucleosome structure on gene expression, what advantages would you see in maintaining such information within the chromatin structure?

D. Standardized test format

1. All of the following may be found within a chromosomal domain except

    A. hypersensitive sites.
    B. promoter elements.
    C. enhancer elements.
    D. a locus control region.
    E. all of the above may be found in a chromosomal domain.

2. The average cutting periodicity of DNAase I directly reflects what characteristic of the DNA helix?

    A. linker DNA length
    B. pitch of the super helix
    C. base pairs per turn of the helix
    D. average length of gene
    E. length of DNA per nucleosome

3. Considering the interaction between DNA and the nucleosome core particle, what kinds of bonds hold the DNA to the core histone octamer?

    A. ionic
    B. hydrophobic
    C. covalent
    D. C only
    E. A and B only

4. Which of the following accurately describes the nucleosome as it exists in a mammalian cell nucleus during the mitotic phase of the cell cycle?

    A. 2(H2A), 2(H2B), 1(H3), 1(H4), ~146 base pairs of DNA
    B. 2(H2A), 2(H2B), 2(H3), 2(H4), ~200 base pairs of DNA
    C. 1(H1), 2(H2A), 2(H2B), 2(H3), 2(H4), ~200 base pairs of DNA
    D. 1(H1), 2(H2A), 2(H2B), 1(H3), 1(H4), ~200 base pairs of DNA
    E. 1(H1), 1(H2A), 1(H2B), 2(H3), 2(H4), ~146 base pairs of DNA

5. The relationship between hypersensitive sites and the expression of genes associated with them is that

    A. DNA sequences contained within hypersensitive sites are required for gene expression.
    B. promoters are inactivated by hypersensitive sites.
    C. all genes throughout the genome maintain promoters containing hypersensitive sites.
    D. hypersensitive sites maintain freedom from nucleosomes at all times.
    E. hypersensitive sites are completely free of protein association.

6. Considering the importance of charge maintained on the N-terminal histone tails, the elimination of this amino acid(s) would result in altered nucleosome function.

    A. lysine
    B. arginine
    C. histidine
    D. A and B only
    E. A, B and C

7. Two terms used to describe the positioning of DNA on the surface of nucleosomes are

    A. transverse and intrinsic.
    B. extrinsic and intrinsic.
    C. periodicity and transverse.
    D. boundary and rotational.
    E. translational and rotational.

8. A cluster of genes whose expression is influenced by a locus control region can be described as a

    A. core particle.
    B. activator.
    C. chromosome.
    D. domain.
    E. element.

9. Experiments designed to identify if core histones are dissociated for reuse or they remain assembled during DNA replication used heavy (^{15}N) and light (^{14}N) isotopes of nitrogen. The histone core particles assembled following DNA replication included which of the following characteristics?

    A. Exclusively light isotopes of nitrogen
    B. Exclusively heavy isotopes of nitrogen
    C. An equimolar mixture of heavy and light isotopes
    D. A and B
    E. A, B and C

10. Considering the need for the reduction of the positive charge of N-terminal histone tails during chromatin remodeling, modifications were observed in which the hydroxyl group of serine was phosphorylated and the amino group of lysine was methylated. How can these modifications be justified?

     A. Methylation and phosphorylation both introduce negative charges.
     B. Positive charge of the hydroxyl is countered by the negative methyl group.
     C. The methyl group eliminates the positive charge of amino and phosphate provides negative charge.
     D. Phosphate eliminates negative of hydroxyl and methyl provide negative charge.
     E. There is no justification for such modifications in chromatin remodeling.

Answers:

A. Recall and short answer

1. The nucleosome provides the first level of chromatin structure. This consists of 146 base pairs of DNA wrapped around a histone core particle providing a packing ratio of ~ 6. A second level of chromatin organization is the 30 nm chromatin fiber, which is a coiled array of nucleosomes consisting of 6 nucleosomes per turn of the fiber. The 30 nm chromatin fiber provides a packing ratio of ~ 40. A third level of chromatin organization is the packaging of the 30 nm coiled fiber into a higher order looped coil structure providing a packing ratio of ~ 1,000. An increased packing ratio value reflects a greater level of structural organization of the chromatin.

2. Euchromatin and heterochromatin exist on the same chromosome throughout the cell cycle. Euchromatin maintains a variable level of organization throughout the cell cycle varying from nucleosome organization with a packing ratio of ~ 6 to the mitotic chromosome with a packing ratio of ~ 10,000. The heterochromatin generally maintains a nucleosome organization with a packing ratio of ~ 10,000 throughout the cell cycle. Lower levels of ultrastructural organization result in expression of DNA, therefore DNA sequences organized into euchromatin are readily expressed, while DNA within heterochromatin is transcriptionally silent. The relative proportion of euchromatin and heterochromatin that comprise a chromosome can change as cells differentiate as a consequence of the need for differential gene expression.

3. The nucleosome core contains 2 copies each of histones H2A, H2B, H3 and H4. Histones H2A and H2B form dimers, while histones H3 and H4 form dimers. The monomers maintain a molar ratio of 1 therefore each monomer is represented equally within the core histone particle.

4. A 200 base pair ladder is observed when all of the chromatin is organized within nucleosomes reflecting a packing ratio of ~ 6.

5. DNA free in solution is nicked at random while nucleosome associated DNA is nicked at regular 10-11 base pair intervals. The product of the enzyme's action on the nucleosome bound DNA reflects the number of nucleotide base pairs per turn of DNA (structural periodicity) within the nucleosome core.

6. core DNA; linker DNA

7. The 10 nm chromatin fiber is a string of nucleosomes connected by linker DNA maintaining a packing ratio of ~ 6. The 30 nm chromatin fiber is a coiled organization of nucleosomes maintaining ~ 6 nucleosomes per turn of the fiber assuming a packing ratio of ~ 40.

8. Yes; the H2A-H2B and H3-H4 dimers each interact with the phosphodiester backbone of DNA and the histone N-terminal tails contact linker DNA and core histones of adjacent nucleosomes.

9. The N-terminal tails of core histones are subject to acetylation, methylation and phosphorylation, which reduces the net positive charges that reside on various amino acids of the histone tails.

10. replication; replication independent. All nucleosome synthesis must not be linked to DNA replication, as other processes including DNA repair require nucleosome disruption and nucleosome synthesis.

11. Intrinsic; extrinsic model; the first nucleosome in a region maintains a sequence specific interaction, which determines a boundary that restricts the positions available for adjacent nucleosome assembly. A series of nucleosomes subsequently forms within the adjacent region.

12. translational positioning; rotational positioning; translational positioning refers to the sequences of DNA that are found within the linker DNA regions and which DNA regions are displayed on the nucleosome core. Linker DNA is accessible to nuclease action suggesting the DNA is accessible to transcription factor binding. Core DNA is accessible to several transcription factors (see steroid receptor binding) while inaccessible to others. Rotational positioning refers to the sequence of DNA that is displayed on the histone core. Displayed DNA can maintain important contact points needed for transcription factor binding.

13. Yes, genes that are being transcribed maintain nucleosome association at the same frequency as nontranscribed genes. However, the nucleosomes are temporarily displaced as the RNA polymerase transcribes the gene segment.

14. Hypersensitive sites reflect an increased sensitivity to the DNAase I enzyme. Hypersensitive sites appear in promoter regions prior to the initiation of transcription. The nucleosomes are excluded from various DNA sequences by proteins, which include the transcription factors required to activate the promoter.

15. The locus control region is a sequence of DNA that maintains a series of hypersensitive sites within a defined DNA domain. The locus control region is absolutely required for the expression of clusters of genes with adjacent DNA sequence, reflecting the association of hypersensitive sites with active gene expression.

B. *Analytical and critical thinking*

1. (a) CAF-1 links the process of DNA replication and nucleosome assembly. The CAF-1 protein complex facilitates the assembly of the nucleosome core by binding histone H3 and H4 monomers. Therefore an *in vitro* experiment examining replication dependent nucleosome assembly in the absence of CAF-1 would yield negative results.

(b) Replication independent nucleosome assembly does not require CAF-1 therefore success may be achieved in the absence of this protein complex.

2. (a) A hypersensitive site reflects the availability of DNA within chromatin. The DNA contained within a hypersensitive site is not organized in nucleosomal structure, therefore it is sensitive to DNAase I action. Many hypersensitive sites are associated with active promoters and the subsequent expression of genes. The hypersensitive site appears prior to transcription and the sites contain DNA sequence that is essential for the initiation of transcription.

(b) The activated promoter sequences maintain boundary regions designated by transcription factor binding, which excludes further association with nucleosomes. The resultant "free" DNA sequence is available for transcription initiation.

(c) Basal transcription factors required for the recruitment and stabilization of an RNA polymerase are likely proteins to target hypersensitive sites.

3. Nucleosome monomers containing at least 165 base pairs of DNA can retain histone H1 but the histone is always lost from the nucleosome containing 146 base pairs of DNA. The physical association suggests that histone H1 is associated outside of the histone core kernel, within the linker DNA adjacent to the histone core. The physical location of histone H1 coupled with the solubility of histone H1 depleted DNA suggests histone H1 functions to "seal" the DNA in the nucleosome.

4. (a) No; regions of core DNA rich in A and T nucleotides locate in such a manner that the minor groove of the DNA faces towards the nucleosome core, while core DNA regions rich in G and C arrange so that the minor groove faces away from the histone core. DNA regions that form extreme structures are excluded from nucleosome core organization. Linker DNA is variable in length and is heterogeneous in sequence. However, the organization of boundaries within linker DNA is rigorously maintained suggesting the sequence specific association with boundary proteins.

(b) Translational and rotational positioning reflects the association of DNA with the histone core. The movement of DNA along the core will reflect a change in linker and core DNA sequence, making new sequence available for protein interaction on the surface of the nucleosome or within the linker DNA. Proteins that readily form boundaries can recognize the DNA made available by repositioning the DNA along the histone core.

(c) The rotational position of DNA within the histone core is repressive to the initiation of transcription in many cases. Regions rich in A and T nucleotides locate within the nucleosome in such a manner that the minor groove of DNA faces the histone core. Access to the minor groove of DNA is required for the proper placement of RNA polymerase during transcription initiation. Therefore, any factor eliminating the reorganization of boundary regions necessary to facilitate minor groove access will inhibit promoter activity.

5. The ionic strength of a solution determines the stability of many protein complexes, including the histone core. Conditions of high ionic strength destabilize the histone core allowing for RNA polymerase movement through the nucleosome and therefore successful transcription in the absence of chromatin remodeling complexes.

C. Extensions and applications

1. (a) Chaperone proteins have the ability to form interactions with other proteins. Chaperone proteins may, amongst other functions, structurally support, refold or sequester other proteins.

(b) The FACT complex appears to both facilitate transcription and promote nucleosome assembly by interacting with the histone dimer H2A-H2B. The FACT complex must allow RNA polymerase access to the nucleosome core DNA by destabilizing the core histone H2A-H2B dimer, thus facilitating transcription. In addition the FACT complex generates high molecular weight chromatin through the use of its chaperone activity on the H2A-H2B histone dimer. The FACT complex must actively bind the H2A-H2B histone dimer and facilitate the formation of the core histone complex.

(c) A model for FACT action must include association with the RNA polymerase where the FACT protein complex actively dissociates histone core particles ahead of the polymerase and subsequently reassembles the core histone following polymerase passage.

2. (a) The "Histone code" contains information regarding what DNA sequences are repressed, silenced or actively expressed. Therefore the physical coded information reflective of the covalent modification on the histone core protein actively directs families of proteins to locations within the genome and dictates their level of association with other available proteins. A given segment of chromatin from different cell types may maintain different covalent modifications (a different "Histone code") as a consequence of cellular differentiation and a need for differential gene expression.

(b) This method of communication between proteins and DNA provides a tremendous advantage when considering the control of gene expression. The system provides the information that will allow for rapid assessment of DNA status and further, provides a mechanism for the rapid alteration of the chromatin status via chromatin remodeling enzymes like demethylases, methylases, acetylases, deactylases and phosphorylases.

D. Standardized test format

1. E
2. C
3. E
4. B
5. A
6. E
7. E
8. D
9. C
10. C

Chapter 21

Promoters and enhancers

General Concepts	Section	Related Questions
Promoters	21.1 - 21.7; 21.14	A1, A2, A3, B1, C2, D1
Basal transcription apparatus	21.4 - 21.10	A1, A4 , A5, A7, D2, D3
Activators / enhancer elements	21.13 - 21.17	A11, A12, D8
Insulator elements	21.20 - 21.23	A13, B3, D6, B5
Processes/structures		
Transcription initiation	21.4 - 21.10	A1, A4, A7, A9, D3
TBP binding	21.8 - 21.9	A4, A5 , A8, D2, D4
Promoter clearance	21.11	A6, C1, D5
Transcription and DNA repair	21.12	A10, C1, D10
Demethylation of DNA	21.18 - 21.19	A13, B2, B4, D7, D9
Insulator activity	21.20 - 21.23	A14, A15, B3, D6
Examples, Applications and Techniques		
DNA footprinting / mutagenesis	21.3	A3, B1
Methylation and endonuclease action	21.18	A13, B2, D9

Questions:

A. Recall and short answer

1. Genes maintain a variety of DNA sequence elements within their core promoters. All three eukaryotic RNA polymerases recognize core promoter elements that reside upstream of a genes start site. However, certain RNA polymerases recognize core promoter elements that reside downstream of a genes start site (internal elements). Name the RNA polymerases that recognize internal promoter elements and list the groups of RNAs that these polymerases produce.

2. Each eukaryotic polymerase is responsible for transcribing unique genetic messages. Correlate the RNA product of the polymerase and physical location of polymerase activity within the nucleus.

3. Experiments have provided direct evidence for the essential nature of particular nucleotide elements within eukaryotic core promoters recognized by each RNA polymerase. Would you expect modification of nucleotides within the promoter region to enhance or reduce transcription efficiency? Explain your reasoning.

4. What is the function of the TATA binding protein?

5. Considering the preferences for the binding of TBP at AT rich regions of DNA, what potential restriction on the initiation of transcription does the rotational position of DNA on the surface of the nucleosome core particle present?

6. Select structural modifications of the CTD tail of RNA polymerase II dramatically reduced the success of transcription initiation. What other events involving gene expression may these structural modification influence? Why?

7. For each RNA polymerase, describe the protein factors that allow the polymerase proper placement at the core promoter.

8. Many transcription factors interact with DNA in a sequence specific fashion. The vast majority of DNA binding proteins physically interact with DNA nucleotides in the _____ _____, while the exception to this rule, TBP, interacts with DNA in the _____ _____.

9. RNA polymerase II recognizes two distinctly different core promoters. What nucleotide sequence elements does each promoter maintain?

10. What is the role of TFIIH during the transcription and the repair of DNA damage?

11. Describe the role of an activator.

12. An enhancer held at a fixed distance from the promoter recognized by RNA polymerase II does not effectively influence the initiation of transcription, however the same enhancer brought into physical contact with the promoter facilitates efficient transcription. From these results deduce a mechanism explaining the enhancer's action.

13. The location of GC islands throughout the genome appears to be a consistent predictor of gene activation and expression. What are some of the characteristics maintained by the chromatin in these regions?

14. Particular nucleotide sequence elements have been characterized within the genome that increases the precision of gene regulation by limiting cross talk between regulatory domains. What are these particular elements called? Describe the function of these elements.

15. Describe the structural elements that compose a chromosomal regulatory domain. How does the organization of the regulatory domain explain the overall size of the genome?

B. Analytical and critical thinking

1. (a) The nucleotide sequence elements within the promoters of genes recognized by RNA polymerase II are assembled in a "mix and match" fashion. Explain the regulatory significance of this organizational pattern and provide an example.

(b) What effect would removal of a nucleotide sequence element by site directed mutagenesis have on promoter activity? Why?

2. Cells grown in the presence of 5-azacytidine (experimental) demonstrate changes in states of differentiation following DNA replication while cells grown in cytidine (control) demonstrate no changes in gene expression following replication. Explain the effect of 5-azacytidine on gene expression. What observable experimental result would occur following DNA replication if cellular DNA methyltransferases were inhibited in both the experimental and control conditions?

3. The introduction of foreign DNA into *Drosophila* is essential for the continued study of gene function in animals. In an attempt to examine the regulation of transcription a researcher inserted a novel gene, flanked by two insulator elements, into a regulatory domain adjacent to a large stretch of heterochromatin. Will this insertion location allow for expression of the novel gene? Why or why not? Would the outcome of the transgenic manipulation differ if the gene is inserted into a mutant cell line lacking the gene encoding BEAF-32? Why or why not?

4. Methylation of specific nucleotide sequence and deacetylation of histone proteins modifies local chromatin structure and inhibits the expression of DNA. Describe a model that includes a direct connection between the two forms of covalent modification for the silencing of gene expression. Would the link between the processes exist if *Sin3* was mutated and dysfunctional within your model? Why or why not?

5. In an attempt to measure insulator strength a continuous segment of DNA encoding the 5 *gbk* genes (5' - *gbk*1-*gbk*5- 3') was constructed. Each gene maintained its own regulatory sequences facilitating the expression of each gene. Replacement of the existing 5' boundary region of *gbk* 1 by the insertion of 3 insulators (In1-In3) produced the following results.

Insulator element	genes expressed
In1	*gbk* 5
In2	*gbk* 2,3,4,5
In3	*gbk* 4,5

Which insulator is the strongest? What do the results suggest regarding the strength of the insulator and the distance that it can act? Do insulators of differing strengths exist in eukaryotes and if so how can this difference be explained?

C. Extension and applications

1. The XPD and XPB subunits of TFIIH maintain helicase activity. Work done by de Boer et al. (2002 *Science* 296:1276-1279) demonstrated that mice with a mutation in the *XPD* gene exhibited premature aging, infertility and various age related infirmities. What is the function of the subunit under normal conditions? In terms of understanding the aging process, what insight does the results of the experiment provide?

2. The eukaryotic RNA polymerases maintain unique locations within the nucleus; recognize unique nucleotide sequence elements and maintain unique component parts. Therefore, it is likely that each polymerase associates with unique proteins that are required to remodel chromatin, allowing for the complete transcription of genes. Work by Saunders et al. (2003 *Science* 301:1094-1096) in *Drosophila* indicates that the chromatin remodeling complex FACT (Facilitates Chromatin Transcription) demonstrates a global distribution and localizes within the nucleolus, however it is conspicuously absent from the genes transcribed by RNA polymerase III. Further studies demonstrate a redistribution and colocalization of FACT at loci encoding *hsp* genes in response to a heat shock stimulus. What genes are transcribed by each of the polymerases? Which classes of genes are transcribed with the help of FACT? Considering the localization results what is the likelihood of FACT involvement in the successful transcription of all genes transcribed by RNA polymerase II. Why?

D. Standardized test format

1. Which of the following accurately describes the relationship between each RNA polymerase and its product?

> A. pol I - mRNA, pol II- tRNA, pol III-rRNA
> B. pol III- rRNA, pol I-tRNA, pol II- mRNA
> C. pol III-rRNA, pol II-mRNA, pol I-tRNA
> D. pol III-tRNA, pol I-rRNA, pol II-mRNA
> E. pol II-mRNA, pol I-tRNA, pol III-rRNA

2. Considering the organization of the core promoter, all three polymerases rely on this positioning factor to ensure recognition and proper placement.

> A. CAAT box
> B. activator
> C. TBP
> D. insulator
> E. TAFs

3. The complex required for recognizing the core promoter and providing a target for RNA polymerase II binding is called

> A. enhancers
> B. basal apparatus
> C. insulator complex
> D. core proteins
> E. assembly apparatus

4. Interaction of TBP within the minor groove of DNA is difficult in the presence of nucleosomes because

    A. the minor groove of DNA faces the surface of the nucleosome core particle.
    B. TAFs are inhibited from binding by nucleosome surface charges.
    C. SL1 binds the nucleosome core blocking TBP binding.
    D. the minor groove is physically distorted by nucleosome association.
    E. TBP is repelled by the positive charges of the core histones.

5. All of the following actions are required for promoter clearance by RNA polymerase II except?

    A. ATP hydrolysis
    B. polymerase CTD phosphorylation
    C. TFIIH methylation
    D. DNA melting
    E. helicase activity of XPB

6. Considering the mechanism of interaction between an enhancer and a promoter, what must an insulator do to uncouple the enhancer influence from the promoter?

    A. induce positive supercoiling of DNA at the promoter
    B. block the physical interaction between the enhancer and promoter
    C. create a single base pair cut in the DNA relaxing the DNA supercoil
    D. induce a reciprocal translocation separating the elements
    E. A and B

7. The relationship between CpG islands and promoter activity is that

    A. CpG islands exist within all active promoters.
    B. Methylated CpG islands increase promoter action.
    C. CpG rich islands are conserved in the promoter regions of many active genes.
    D. CpG islands inhibit enhancer action.
    E. A and C

8. Considering that enhancers and promoters contain the same sequence elements, the unique function and significance of the enhancer is that

    A. the enhancer inhibits promoter activation by competing with the promoter for factors.
    B. the enhancer brings a great deal of protein factors in proximity to the basal complex.
    C. the enhancer insulates the promoter from other elements controlling activation.
    D. the enhancer changes supercoiling of the DNA at the promoter.
    E. B and C

9. Undermethylation of the C residues of selected promoter regions appears to coincide with an increased level of promoter activity. Characteristics of an undermethylated promoter include

    A. increased sensitivity to DNAase I.
    B. increased cleavage of the promoter sequence by HpaII.
    C. rearranged chromatin structure.
    D. A and B
    E. A, B and C

10. The processes of DNA repair and the initiation of transcription by RNA polymerase II are linked by the actions of which protein factor?

    A. TFIIJ
    B. TFIID
    C. TFIIH
    D. B and C
    E. processes are not linked

Answers:

A. Recall and short answer

1. RNA polymerases II and III; RNA polymerase II synthesizes mRNAs and RNA polymerase III synthesizes some rRNAs and tRNAs

2. Each RNA polymerase functions within the nucleus. This feature ensures proper RNA processing prior to movement of the RNA species into the cytoplasm. However, the unique localization of RNA polymerase I to the nucleolus coincides with its synthesis of ribosomal RNA.

3. Modification of nucleotide sequence within the core promoter region may be benign or deleterious to successful transcription. Highly conserved nucleotide sequence elements ensure proper placement and binding of RNA polymerase at the start site of the gene. Disruption of RNA polymerase placement can result in reduced copy number of the transcript or an incomplete transcription where no transcript is produced.

4. Presented in several ways as part of the basal transcriptional apparatus, the TATA binding protein is a highly conserved structural protein component of the positioning factor. The positioning factor is necessary for the effective binding and placement of RNA polymerase within the core promoter.

5. The TATA binding protein interacts with AT rich regions within the minor groove of the DNA double helix. Without the modification of the rotational positioning of the double helix the minor groove of the AT rich DNA sequence may not be available.

6. RNA processing; The mRNA 5' capping enzyme (Guanylyl transferase) binds the phosphorylated CTD tail of RNA polymerase II. The SCAF proteins, which bind mRNA splicing factors, also interact with the CTD tail as well as components required to add the polyadenine tail to the 3' end of the mRNA. These protein associations suggest that the CTD tail of RNA polymerase II influences mRNA processing.

7. TFIIIB for RNA polymerase III
    TFIID for RNA polymerase II
    SL1 for RNA polymerase I

8. major groove of DNA; minor groove of DNA

9. TATA containing promoters include the TATA box and the initiator sequence; TATA less promoters include the initiator sequence and the downstream promoter element (DPE)

10. The helicase activity of the XPD subunit of TFIIH is responsible for melting DNA during transcription initiation and DNA repair.

11. These upstream nucleotide sequence elements of the promoter regulate the efficiency and specificity with which the promoter is recognized by RNA polymerase II.

12. Enhancer elements effectively influence the initiation of transcription by increasing the concentration of activator proteins in close physical proximity of the core promoter.

13. Regions of chromatin that contain GC islands maintain changes that are associated with gene expression. This chromatin maintains reduced histone H1 content, acetylated core histones and demonstrates increased sensitivity to DNAase I.

14. Insulator sequence elements bind proteins, which limit interaction between adjacent regulatory domains. The insulator elements and their binding proteins effectively provide a barrier that limits the propagation of heterochromatin and the activating effects of adjacent enhancers.

15. A regulatory domain contains an insulator, a locus control region, a matrix attachment site and transcription units. This organization requires a large number of nucleotides and the physical space required for the rearrangement of chromatin structure. Consider that the chromatin containing the transcription units must decondense to allow for promoter activation and gene expression. The large size of the genome would be expected assuming all expressed genes are organized in similar regulatory domains.

B. Analytical and critical thinking

1. (a) A variety of elements can contribute to the function of the promoter however none of the nucleotide sequence elements are essential for all promoters. Therefore, promoters may contain the same sequence elements or maintain a completely different organization. This organizational pattern confers a level of regulation due to the unique combination of nucleotide sequence elements contained within each promoter. The diverse combination of sequence elements ensures activation of the gene will only occur in the presence of a specific combination of DNA binding proteins. The combination of the Oct-1 and Oct-2 activator proteins and Oct nucleotide sequence element in various cell types ensures appropriate gene activation.

(b) The removal of nucleotide sequence elements from the promoter region by site directed mutagenesis would reduce the cell's ability to activate the expression of the particular gene. The nucleotide sequence elements of the promoter help initiate gene expression by recruiting proteins to the vicinity of the promoter establishing the foundation required for RNA polymerase association and initiation of transcription.

2. (a) 5-azacytidine is incorporated into DNA in place of cytidine where it provides a nucleotide location that is resistant to methylation by cellular methyltransferases. The presence of nonmethylated sites following DNA replication leads to an increase in the activity of genes associated with these regions resulting in changes in the pattern of cellular gene expression.

(b) If cellular methyltransferases were inhibited, methylation would not occur following DNA replication resulting in the elimination of methylated cytidine residues in daughter DNA. This enzymatic change would have little consequence in the experimental culture that were grown in the presence of 5-azacytidine however, the control cultures would demonstrate a pronounced change in gene expression as they would be unable to maintain methylated cytidine residues throughout their genome. The experimental and control cultures would demonstrate very similar states of differentiation in the absence of cellular methyltransferases.

3. (a) Yes; the insulator domain will inhibit the propagation of adjacent heterochromatin from incorporating the nucleotide segment containing the novel gene, thus, allowing for gene expression.

(b) The novel gene will almost certainly be incorporated in the adjacent heterochromatin if the proteins serving to organize the insulator complex are missing. The protein BEAF-32 is a common component of the insulating complex and its absence will likely result in structural changes throughout the genome.

4. (a) Methylated DNA facilitates the association of auxiliary proteins that act to repress gene expression. These proteins, including (MeCP1/2) recognize methylated DNA and directly repress the initiation of transcription by interacting with promoter specific proteins. In addition, MeCP2 recruits Sin3, which contains histone deacetylase action, further repressing activity of the promoter sequence. Therefore, DNA methylation and histone deacetylation can be combined to efficiently silence gene expression.

(b) No; disruption of Sin3 action would result in the loss of deacetylase action within the repressor complex.

5. The insulator In1 maintained the strongest action as it suppressed the expression of four of the five transcription units. The results suggest that the stronger the insulator the further its effects from its position. Yes, insulators of different strengths exist in eukaryotes however the mechanism by which the strength of an insulator is measured is not completely understood. The effect appears to be more than a chromatin-remodeling phenomenon as competitive effects between the insulators are observed.

C. Extension and applications

1. (a) The XPD subunit of TFIIH is responsible for the formation of the initial transcription bubble and melting damaged DNA within various DNA repair mechanisms.

(b) The results of the study strongly suggest the inability to repair DNA leads to cellular decay and accelerated aging. In addition, the results link reduced global transcription with a disease phenotype. The study further implies that any mutation limiting DNA repair or reducing global transcription will limit the organisms ability to effective maintain cellular processes.

2.   RNA polymerase III synthesizes tRNA
     RNA polymerase II synthesizes mRNA
     RNA polymerase I synthesizes rRNA

(a) The FACT complex assists in the transcription of mRNA and rRNA, while it is conspicuously absent from promoters recognized by RNA polymerase III.

(b) It is highly likely that the FACT complex assists in the transcription of all genes recognized by RNA polymerase II as the protein is globally distributed under normal circumstances and redistributes to specific promoter regions during environmental stress, such as heat shock. The heat shock genes are transcribed by RNA polymerase II demonstrating FACT association with RNA polymerase II under this circumstance.

D. Standardized test format

1. D
2. C
3. B
4. A
5. C
6. B
7. C
8. B
9. E
10. C

Chapter 22

Activating transcription

General Concepts	Section	Related Questions
Transcription factors	22.1 - 22.3	A1, A2, A3, A4, D1,
Nucleotide response elements	22.7	A9, D2, D8
DNA binding domains	22.8 - 22.16	A11, A12, A13, B1, D5,
Protein dimerization domains	22.11 - 22.16	A14, A15, B5, C1, D7,
Processes/structures		
Activator action	22.5 - 22.7; 22.13; 22.15 - 22.16	A6, A7, A8, A10, A15, C1, D3, D4,
Activator binding	22.8 - 22.9; 22.13; 22.14 - 22.16	A11, A12, A13, A14, B1, B4, D5, D6, D9,
Combinatorial association	22.13 - 22.16	A14, B5, D10
Examples, Applications and Techniques		
Yeast two hybrid assay	22.4	A5, C2

Questions:

A. Recall and short answer

1. The initiation of transcription requires extensive protein-protein interaction. Many promoters require _____ proteins to bridge the gap between proteins bound to the basal promoter and those bound at activator elements.

2. Unlike bacterial RNA polymerase, the eukaryotic RNA polymerase II does not physically interact with all protein factors required for initiation of transcription. Briefly describe the group(s) of initiation factors that do not directly interact with the polymerase.

3. Many proteins involved in the process of transcription initiation maintain two functional domains. What are functional domains and how do they facilitate protein function?

4. Many DNA binding proteins maintain domains that recognize and interact with nucleotide sequence elements in a _____ specific fashion. The numerous weak interactions that form between the _____ R groups of the protein domain and the nucleotides within the _____ of the alpha helix are responsible for this recognition.

5. A technique used to ascertain the function of protein binding domains is the two-hybrid assay. Briefly describe the two hybrid proteins generated within this technique.

6. What protein-protein interactions occur between activators and coactivator proteins to facilitate the initiation of transcription for RNA polymerase II?

7. Molecular charge is a crucial factor in facilitating the productive interaction of various proteins, including activator proteins, within the initiation complex of RNA polymerase II. The net negative charge of an acidic activator domain facilitates what process?

8. An experiment conducted to test the function of activators on the initiation of transcription demonstrated that activator action was of little consequence at elevated polymerase concentrations. What conclusion could be drawn from the same experiment if activator action was required at very high and very low polymerase concentrations?

9. Name the nucleotide sequence elements that are recognized by activators, may be found within the promoter or enhancer regions of widely distributed genes and will confer common transcriptional control over groups of similar genes.

10. Is it possible that a nucleotide sequence element needed for the activation of a gene in one particular cell type is involved in the repression of the same gene in another cell? Briefly explain.

11. Activator proteins exist in many forms and the availability of these proteins imparts a level of regulation on gene activation. Explain how effective dimer formation of activator monomers regulates the activation of particular promoters.

12. Zinc finger motifs are named for the metal ion that coordinates the structural motif. However, this feature of the motif recognizes and binds DNA in a sequence specific fashion. Describe the DNA binding domain contained within the zinc finger motif and give an example of an amino acid contained within the zinc finger motif that would effectively interact with DNA.

13. Mutations caused by DNA adduct formation changed the two amino acids between the coordinating cysteines on the right side of the first zinc finger of the glucocorticoid receptor. Phenotypic results demonstrate the receptor now has marginal affinity for the estrogen response element. Is this result a possibility? Explain.

14. DNA binding proteins maintain physical characteristics which allow for the sequence specific recognition and binding of nucleotide sequence elements and also maintain the tertiary structure enabling effective dimer formation. Describe the classes of amino acid responsible for each interaction.

15. The combinatorial association of various protein monomers is a form of transcription regulation. This form of regulation is typified in describing the interaction between proteins maintaining HLH and bHLH motifs. Briefly explain.

B. Analytical and critical thinking

1. Many DNA binding proteins maintain DNA binding and protein dimerization domains. What amino acids compose the DNA binding and the protein dimerization domain? What noncovalent interactions are formed between the nucleotides within the major groove of DNA and the amino acids of a recognition helix? Why are basic regions common components of DNA binding domains?

2. Steroids control gene expression via the combinatorial association of monomers resulting in effective DNA binding and promoter activation. The activation of gene expression by steroids occurs by the recruitment of coactivator complexes and the remodeling of chromatin. What is a coactivator and how does steroid receptor binding to the response element result in the remodeling of chromatin and the initiation of gene expression? What effect on the expression of the target gene would be expected if the CpG islands within the promoter region of the gene were fully methylated? Why?

3. A family of HLH proteins forms homo and heterodimers to effectively control selective gene expression. The family of activator proteins includes 8 members maintaining basic regions adjacent to the HLH domain (bHLH) and two members lacking the basic region. What function does the basic region maintain? The formation of homo or heterodimers between the bHLH family members resulted in effective activation of gene expression while the formation of heterodimers between a bHLH family member and nonbasic HLH family member failed to activate gene expression. Why is this so? Describe how the combinatorial association of the family members regulates monomer availability and the subsequent activation of gene expression.

4. Cholesterol metabolism is tightly regulated in animals, where elevated levels of bile acid inhibit bile synthesis from cholesterol. Within the process, the bile acid receptor (FXR) forms a heterodimer with retinoic acid (RXR) and induces the expression of a strong transcriptional repressor, inhibiting bile acid synthesis. Recent work by Urizar et al. (2002 *Science* 296:1703-1706) demonstrated the plant sterol guggulsterone disrupts this signaling allowing for elevated levels of bile acid synthesis and secretion. The plant sterol does not inhibit receptor dimerization or DNA binding by existing dimer. However, a titration of guggulsterone demonstrated a dose dependent inhibition of gene expression by the sterol. How does the plant sterol inhibit the action of the bile acid steroid receptor? Explain.

5. Recent work by Newman and Keating (2003 *Science* 300: 2097-2101) supports many of the reported partnership interactions between protein monomers containing the bZIP DNA binding motif. Charged and polar amino acid residues at the interface of the monomers are known to contribute to partnering specificity. Describe the dimerization domains of these monomers. What advantage would the charged and polar amino acids residues provide for the effective partnering of these protein monomers?

C. Extensions and applications

1. Molecular chaperones are proteins that have been linked to the rapid inhibition of steroid hormone receptor transcriptional complex action. Work by Freeman and Yamamoto (2002 *Science* 296:2232-2235) demonstrated that preincubation of the protein p23 with DNA effectively eliminated the association of the TR/RXR steroid heterodimer at the thyroid response element. Further analysis demonstrated p23 stopped transcriptional activation of preformed steroid receptor activation complexes in vitro and p23 was selectively recruited to active glucocorticoid response elements in rat liver cells. Based upon these results is p23 a molecular chaperone? Explain. Considering that the rates of transcription of steroid inducible genes change quickly upon hormone exposure and withdrawal what mechanism does the chaperone protein use to facilitate these changes in gene expression?

2. Huntington's disease is a dominantly inherited neurodegenerative disorder presenting neurologic and muscular impairment leading to death. The mutant protein Huntington maintains an expanded polyglutamine tract with which it maintains unusual protein/protein interactions. Work by Dunah et al. (2002 *Science* 296: 2238-2242) demonstrated the disorder is characterized by the disruption of Sp1/TAFII130 activator function. The yeast two-hybrid technique demonstrated that the mutant Huntington protein interacts with both the ubiquitous transcription activator Sp1 and the transcriptional coactivator TAFII130. Propose a mechanism for the Huntington protein within the diseased state based upon these results.

D. Standardized test format

1. Which of the following effectively bind DNA and facilitate gene expression?

    A. basal factors
    B. coactivators
    C. activators
    D. A and C
    E. A and B

2. Response element function usually involves a situation in which

    A. gene silencing occurs due to basal apparatus destruction.
    B. proteins bind to a fixed location within the promoter.
    C. covalent modification of chromatin is inhibited.
    D. RNA polymerase is bound and removed from the basal apparatus.
    E. activators bind the promoter or enhancer.

3. Acidic activators are effective in facilitating

    A. RNA polymerase clearance from the basal apparatus.
    B. TFIIB introduction into the basal apparatus.
    C. modification of the CTD tail of RNA polymerase.
    D. the stabilization of TAFs within TFIID.
    E. A and B

4. The repression of transcription in eukaryotes is usually accomplished by

    A. repressor proteins binding RNA polymerase II.
    B. repressor proteins binding TBP.
    C. repressor proteins binding nucleosomes and influencing chromatin structure.
    D. A and C
    E. A, B and C

5. Considering that inducible activators of gene expression maintain many structural differences, their actions may be regulated in all of the following ways except

    A. ligand binding to the activator.
    B. direct covalent modification of the activator (phosphorylation).
    C. combinatorial dimerization of monomers.
    D. enzymatic modification of the activator protein (cleavage).
    E. All of the actions above may regulate activator action.

6. The sequence specific interaction between DNA binding motifs of activators and nucleotides within the major groove of DNA occurs through the use of an alpha helix in all of the following except

    A. zinc finger motif.
    B. basic helix loop helix motif.
    C. leucine zipper motif.
    D. homeodomain.
    E. A and D

7. The formation of effective dimers is essential for nucleotide sequence specific recognition and binding for which of the following DNA binding motifs?

    A. homeodomains
    B. leucine zippers
    C. steroid receptors
    D. A, B and C
    E. A and C

8. Consider an insertion mutation that alters the signal activating the expression of a particular gene from the 9-cis-retenoic acid homodimer to the retinoic acid-vitamin D heterodimer. What is the most likely explanation for this observation?

    A. Mutation decreased the spacing between half elements in the response element
    B. Mutation eliminated one half element eliminating the need for dimer formation
    C. Vitamin D altered chromatin structure eliminating the need for receptor binding
    D. Mutation increased the spacing between the half elements in the response element
    E. There is no logical explanation for the observation

9. DNA binding motifs common to activators make the majority of their sequence specific interactions with DNA using which of the following?

    A. an alpha helix.
    B. a region of basic amino acids.
    C. a beta sheet structure.
    D. a coordinated metal ion.
    E. A and B

10. Considering that in a developing muscle cell the bHLH heterodimer MyoD-E12 is a trigger for myogenesis, overexpression of the nonbasic HLH monomer Id in these same cells prevents myogenesis by

     A. creating a new effective dimer Id-E12 which activates different genes.
     B. creating a trimer with MyoD-E12 that is unable to bind DNA.
     C. effectively displacing the E12 monomer creating an ineffective MyoD-Id dimer.
     D. sequestering E12 rendering the monomer unavailable for MyoD-E12 formation.
     E. A and D

Answers:

A. Recall and short answer

1. coactivators

2. Activators and coactivators do not interact directly with RNA polymerase II. Activators may interact directly with the basal apparatus or interact with the basal apparatus through a coactivator.

3. The functional domains include a DNA binding domain and a transcriptional activation domain. The DNA binding domain brings the protein into the vicinity of the promoter and the transcription activation domain facilitates the interaction between different proteins within the initiation complex.

4. sequence; amino acid; major groove

5. The technique tests whether the two protein domains can interact with each other. The first hybrid protein contains the first experimental domain fused to a DNA binding domain that will bind a promoter facilitating the expression of a reporter gene. However, the first hybrid protein will not initiate gene expression unless it interacts with a second hybrid protein. The second hybrid protein contains a transcription activation domain fused to the second experimental domain. The two experimental domains must effectively interact brining the two proteins into close proximity, thus facilitating the initiation of gene expression.

6. Coactivators make many protein-protein interactions with members of the basal apparatus. The coactivator proteins may target TFIID, TFIIB or TFIIA as all of these factors participate in the early stages of basal apparatus assembly.

7. An acidic activator protein maintains a domain that contains a series of negative charges. The charged motif facilitates the proteins interaction with DNA and other proteins. In particular, the acidic activator targets TFIIB, increasing the rate at which it binds the initiation complex. Since the recruitment of TFIIB is rate limiting, the stimulation of TFIIB association increases the rate of transcription initiation.

8. Activator proteins enhance the rate at which RNA polymerase will bind the initiation complex. Therefore, if activator action did not enhance the rate of transcription at elevated RNA polymerase concentrations the most obvious conclusion to draw is that activator action is dependent upon polymerase concentration. Under elevated polymerase concentrations the recruitment of the polymerase to the initiation complex is no longer a rate-limiting event. If the same experiment resulted in activator action at all concentrations of polymerase it would suggest that polymerase recruitment into the initiation complex was rate limiting regardless of concentration.

9. response elements

10. Yes; nucleotide sequence elements can be bound by activators in the presence and or absence of particular signals. Binding in the presence of signals can activate the promoter, while activator binding in the absence of signal can result in the recruitment of a corepressor complex silencing activity of the gene. This differential activation can be seen in different cells or in the same cells at different times. The differential activation is typical of steroid receptors and their sequence elements. Particular steroid receptors, in the presence of steroid ligand, will act as an activator while the steroid receptor in the absence of ligand will bind the same sequence elements and facilitate the repression of gene expression.

11. Nucleotide sequence elements are read in a sequence specific fashion by activator motifs maintained in hetero and homo dimer association. Effective dimerization places DNA binding motifs in a position to associate with DNA in a productive fashion, facilitating dimer stabilization and transcription initiation complex formation. The formation of ineffective dimers results in the nonproductive interaction with DNA and the failure to form the transcription initiation complex. The production of effective dimers results in the activation of specific promoters and the regulation of gene expression. The appropriate monomers must be present for effective dimers to be produced consequently a restriction on the availability of monomer type will limit activator action and the resulting activation of gene expression.

12. The DNA binding motif that is formed by the zinc finger is an alpha helix. Two amino acids within the alpha helix make sequence specific contacts with the DNA within the major groove. The amino acids, which effectively interact with DNA, are generally those that maintain basic amino R groups. The basic amino acids maintain a net positive charge at physiological pH allowing the R groups to effectively form hydrogen bonds with the nucleotide base pairs within the major groove. The amino acids within this group include: histidine, lysine and arginine.

13. Yes, the two amino acids in question determine the specificity of the zinc finger for the nucleotide sequence element that is recognized. The two new amino acids must maintain R group chemistry allowing for marginal recognition of the estrogen response element.

14. The amino acid R groups recognized as being able to bind DNA in a sequence specific fashion are those that are able to form hydrogen bonds with nucleotides pairs within the major groove of DNA. The amino acids that facilitate dimerization between protein monomers are those that are stabilized within a hydrophobic environment. Amino acids with charged R groups and those maintaining hydrogen bonding capability can exist in this environment if they are partnered with a like amino acid R group.

15. Effective dimer formation is a crucial factor in the activation of specific gene expression. Many monomers maintain basic amino acids patches adjacent to their dimerization domains, which effectively interact with nucleotide sequence. Effective dimer formation places two basic patches within consecutive turns of the major groove resulting in effective DNA binding. If the nucleotide sequence matches the dimer specificity the formation of the initiation complex will be favored. However, the formation of a dimer in which only one monomer maintains a basic patch results in ineffective DNA binding and does not favor transcription initiation. It is likely that the existence of monomers without basic patches is a form of regulation by which the ineffective monomers bind and functionally eliminate the effective monomers.

B. Analytical and critical thinking

1. (a) The basic amino acids compose the DNA binding motifs of the coactivator proteins. While the uncharged hydrophobic amino acids compose the dimerization domains. (b) The basic amino acid R groups of the DNA binding motifs form many noncovalent interactions with the nucleotide base pairs residing within the major groove of DNA. However, the noncovalent interactions that are typically sequence specific are the hydrogen bonds. (c) The basic region is a common component of the DNA binding domain because it enables the activator protein to interact with DNA in a sequence specific fashion through the formation of hydrogen bonds between the amino acid R groups that reside within the basic patch and the nucleotide base pairs within the major groove.

2. Recall that steroid receptors are activators that bind steroid response elements in the promoter region of the target gene. Activators bind response element sequences in a temporal or developmental fashion controlling patterns of gene expression. Many activators require the recruitment of coactivators, which bridge the physical gap between the activator and the basal apparatus. Steroid receptor binding recruits the coactivator complex P300/CPB, which is known to maintain histone acetylase action. The acetylation of histones changes chromatin structure resulting in the remodeling of chromatin and the activation of target gene expression. Methylation of the CpG islands within the promoter region of the target gene would inhibit steroid binding and silence the expression of the target gene. Recall that methylation of the CpG islands recruit proteins including MeCP2 and Sin3 that deacetylate histones within the promoter region of the target gene resulting in gene silencing.

3. The basic region of the HLH motif binds DNA. The effective dimerization resulted in the formation of aprotein complex that maintained two basic domains, which provides the dimer sufficient affinity to bind DNA and stabilize the coactivator complex. The stabilization of the coactivator dimer allowed for the stimulation of transcription initiation. The combinatorial association of the monomers reflects the interaction of each of the monomers with each member of its protein family. Productive dimer formation occurred between family members that maintained basic domains while nonproductive dimers included at least one family member that lacked the basic region. The formation and maintenance of nonproductive dimers would sequester the bHLH protein from forming productive dimers with other family members and activating gene expression.

4. The plant sterol influences the action of the steroid response element in a dose dependent fashion yet doesn't influence the dimerization of the receptor in the presence of high levels of bile acid or DNA binding by the bile acid associated receptor dimer. Therefore, the plant sterol must be a competitive inhibitor of bile acid, competing with bile acid for receptor binding. The plant sterol must inactivate the receptor once it is bound or generate an ineffective dimer thereby reducing the effective pool of steroid receptor available for bile acid to bind.

5. The bZIP motifs are the highly basic DNA binding domains that reside on proteins that maintain leucine zipper dimerization domains. Leucine zippers present hydrophobic faces that facilitate dimer formation. The hydrophobic nature of the leucine zippers provides little selectivity for perspective dimerization partners. Therefore charged and polar amino acid residues adjacent to the leucine zipper would allow monomers to produce effective dimers based upon common patterns of complementary charges. This charge distribution would provide selectivity and specificity to the dimerization of activators maintaining leucine zipper dimerization domains.

C. Extensions and applications

1. Yes, a chaperone protein may, amongst other functions, structurally support, refold or sequester other proteins. In this example p23 inhibits the production of effective steroid hormone receptor complexes and promotes the disassembly of existing steroid receptor activation complexes by forming noncovalent associations with the component parts. Considering that the rates of transcription of steroid inducible genes increases within 5 to 10 minutes and the cessation of transcription occurs promptly with hormone withdrawal the molecular chaperone p23 actively and reversibly disassembles steroid hormone receptor complexes. The continual disassembly of steroid receptor complexes by molecular chaperones would allow for the fine control of gene expression in response to fluctuations in steroid hormone levels.

2. The expanded polyglutamine tract within the mutant Huntington protein facilitates an aberrant interaction between the Huntington protein and proteins involved in transcription initiation. The yeast two-hybrid technique demonstrated the interaction between the polyglutamine tract of the mutant Huntington protein with both the Sp1 and TAFII130 transcriptional activator proteins. The transcriptional activator Sp1 is a ubiquitous transcription factor that recruits TFIID to the basal transcription apparatus. The transcriptional coactivator TAFII130 interacts with various transcription activators including Sp1 and the cAMP response element binding protein (CREB) implying a role in transcription activation. Tissues expressing the mutant Huntington protein display transcriptional dysregulation due to aberrant protein-protein interactions involving the mutant Huntington protein and transcriptional activator proteins, resulting in a disease phenotype.

D. Standardized test format

1. D
2. E
3. B
4. C
5. E
6. B
7. D
8. D
9. E
10. C

Chapter 23

Controlling chromatin structure

General Concepts	Section	Related Questions
Chromatin	23.1-23.4	A5, A15, D1, D5
Histone core	23.9-23.10	A6, B2, D4
Heterochromatin	23.13-23.17	A12, B1, C1, D3, D6, D8
DNA /alternative states	23.9; 23.19-23.22	A10, A13, A14, B5, D10
Processes/structures		
Chromatin remodeling	23.3-23.5; 23.10	A1, A2, A3, A4, B3, D2
Histone acetylation / deacetylation	23.6-23.8	A6, A8, A9, B4
DNA phosphorylation / methylation	23.9; 23.19-20	A7, A10, A13, B5, C1
Heterochromatin nucleation / propagation	23.13-23.17	A11, B1, B4, C2, D7, D9
Examples, Applications and Techniques		
DNAase I action	23.3	A4
DNA methylation	23.19	A13, B5
RNA inhibition	23.13	C2

Questions:

A. Recall and short answer

1. Can transcription factors bind DNA that is associated with histone proteins? What is the relationship between transcription factor binding and chromatin remodeling?

2. Successful chromatin remodeling is reflected in the increased susceptibility of DNA to DNAase I action. Why?

3. Chromatin remodeling is a process undertaken by large protein complexes that maintain ATPase action. What structural changes within the modified chromatin are dependent upon ATPase action?

4. The introduction of remodeling complexes to chromatin, typified by NURF binding in *Drosophila*, is sequence specific. What action facilitates NURF binding and what changes to chromatin structure result from NURF binding?

5. Nucleosomes do not by their nature exclusively inhibit gene expression. Briefly describe a situation in which nucleosomes display sequences of DNA on their surface in a rotational position that is essential for recognition by transcription factors.

6. A strict correlation does not exist between covalent histone modification and gene activity however,_____ of histones H3 and H4 is associated with active chromatin, while _____ of Lys 9 of histone H3 is associated with inactive chromatin.

7. What function if any does the phosphorylation of histone H3 play within the process of chromatin remodeling?

8. Exposing cells to butyric acid resulted in inhibition of _____, which provided the first observable correlation between acetylation and the _____of gene expression.

9. Describe two functions of coactivator CBP within the transcription initiation complex of RNA polymerase II.

10. The methylation of both histone protein and DNA is associated with DNA inactivity. Considering that some histone methyltransferase enzymes recognize and bind methylated DNA sequence, what does this relationship suggest is responsible for DNA silencing? Why?

11. Heterochromatin maintains highly condensed nucleosomes rendering the genes contained within the region transcriptionally silent. DNA inserted into these regions becomes heterochromatin suggesting an active silencing mechanism, however the heterochromatin does not spread uncontrolled throughout the genome. What protein triggers the production of heterochromatin in mammals and what factors determine boundary limits?

12. How does the inactive X chromosome in human females reflect the need for regulation of gene dosage? Can dosage compensation take place in the absence of DNA methylation? Why or why not?

13. The conservation of DNA methylation following DNA replication is strict considering the role of methylated CG islands within the process of transcription. Briefly describe the relationship between DNA methylation and the existence of hypersensitive sites throughout the genome.

14. Epigenetic inheritance is characterized by the covalent modification of various components within nucleosomes. Methylation of DNA may account for epigenetic inheritance as long as the _____enzyme is active following DNA replication, while the _____of histones H3 and H4 may provide a second epigenetic control mechanism.

15. Describe the physical characteristics that are typical of actively transcribed DNA. Compare and contrast the physical characteristics of actively and poorly expressed DNA.

B. Analytical and critical thinking

1. Chromatin structure and hence gene expression differ in cultured mammalian cells residing in the G0 versus the G1 phase of the cell cycle. Work by Ogawa et al. (2002 *Science* 296:1132-1136) demonstrated the phase specific expression of a repressor complex E2F6.com-1 that created transcriptionally inactive chromatin. The complex contains two proteins maintaining SET domains and several proteins maintaining chromo domains. Provide a model in which E2F6.com-1 acts as a recruitment factor for the repression of gene expression. What happens to the function of the complex if the SET domains were eliminated? Why?

2. In eukaryotes, epigenetic information is contained within chromatin structure. Therefore in many genes, nucleosome positioning must be maintained following transcription. Previous work demonstrated that the chaperone protein complex (FACT – facilitates chromatin transcription) was required for transcription elongation, DNA repair and DNA replication. Recent work by Belotserkovskaya et al. (2003 *Science* 301:1090-1093) demonstrated that FACT caused an increase in dissociated H2A-H2B dimer from the histone core and in subsequent studies demonstrated the complexes ability to generate high molecular weight chromatin, while maintaining rotational and translational positioning of DNA. What is the function of the chaperone protein? How does FACT facilitate the conservation of epigenetic information following transcription?

3. Eukaryotic chromatin structure is modulated by ATP dependent protein complexes including SWI/SNF and NURF. Shen et al. (2003 *Science* 299:112-114) demonstrated that chromatin remodeling occurred in the presence of inositol tetrakisphosphate (IP4) while chromatin remodeling was inhibited in the presence of inositol hexakisphosphate (IP6). What does this relationship tell us regarding the potential signaling mechanisms for the regulation of chromatin remodeling?

4. The molecular mechanism by which particular genes are expressed in certain cells and repressed in others remains a central issue in developmental and regulatory biology. Lunyak et al. (2002 *Science* 298:1747-1752) demonstrated the silencing of neuronal genes in non-neuronal tissues required the repressor protein REST/NRSF. The REST/NRSF protein establishes interactions with mSin3 (HDAC1-2) and recruits the HP-1 protein; what mechanism does the repressor element use to turn off gene expression? Considering the interaction with the HP-1 protein; does the complex also facilitate long term gene silencing? Why or why not?

5. Human cancer cells are characterized by abnormal gene expression. Work by Gaudet et al. (2003 *Science* 300:489-492) demonstrated that transgenic mice displaying 1/10th of the DNA methyltransferase action developed aggressive T-cell lymphomas. What can be concluded regarding these results? Would you expect this phenotype to be seen in mutant mice lacking histone methyltransferase activity? Why or why not?

C. Extensions and applications

1. Genomic imprinting is a form of gene silencing that is epigenetic in origin. The epigenetic effect may differ between individual cells in an animal producing position effect variegation. Work by Cui et al. (2003 *Science* 299:1753-1755) indicates that the loss of imprinting of the gene encoding insulin-like growth factor II is cell specific and is an excellent predictor of cancers, including ovarian, lung, liver and colon. What physical features reflect imprinted DNA? How does the loss of imprinting correlate with increased gene expression and a higher incidence of cancer?

2. RNA silencing (RNAi) is a sequence specific process in which target RNAs are destroyed and DNA sequences are silenced. The process is precipitated by small RNAs, which recruit groups of proteins to facilitate the process. Work by Volpe et al. (2002 *Science* 297:1833-1837) demonstrated in mutant yeast that the deletion of genes encoding the enzymes argonaute, ribonuclease III (dicer) and RNA-dependent RNA polymerase allowed accumulation of RNA transcripts from centromeric heterochromatic repeats and loss of histone H3 lysine-9 methylation from nucleosomes maintaining the centromeric heterochromatic DNA sequences. Complementary work by Zilberman et al. (2003 *Science* 299:716-719) report the argonaute family of proteins in *Arabidopsis* is linked with histone and DNA methylation. What effect does histone and DNA methylation have on gene expression? What factors account for the transcription of DNA contained within the centromere DNA in the mutant yeast system? Does RNAi direct mammalian X inactivation? Why or why not?

D. Standardized test format

1. Which of the following characteristics are typical changes observed during chromatin remodeling?

      A. nucleosome octamer displacement from DNA
      B. nucleosome octamer movement along the DNA
      C. dissociation of the H3 dimer from the core histone kernel
      D. elimination of covalent association between core histones and DNA
      E. A and B

2. Chromatin remodeling complexes require energy to perform which of the following functions?

      A. covalently modify amino acids residing on the tails of histones H3 and H4
      B. unzip the DNA double helix generating single stranded sequence
      C. twist DNA on the nucleosome surface
      D. dissociate the nucleosomes octamer
      E. C and D

3. The relationship between heterochromatin and the histone core is that

      A. histones within heterochromatin maintain acetylations.
      B. histones within heterochromatin contain methylations.
      C. histones within heterochromatin maintain phosphorylation.
      D. histone H3 within heterochromatin lacks an N-terminal tail.
      E. B and C

4. Transcriptionally repressed chromatin normally maintains

    A. acetylation of H3 and H4 histone tails.
    B. methylation of DNA and histone proteins.
    C. histone phosphorylation.
    D. HP-1 association.
    E. B and D

5. Considering that the activation of a promoter involves an ordered series of events, of the actions listed below this action must occur first.

    A. RNA polymerase binds
    B. target DNA sequence is bound by transcription factor
    C. mediator complex binds
    D. histone acetylase complex binds
    E. remodeling complex binds

6. This protein motif targets activating and repressing proteins to specific portions of chromatin.

    A. zinc finger motif
    B. basic helix loop helix motif
    C. chromo domain
    D. homeodomain
    E. A and D

7. This factor likely regulates the spread of heterochromatin from its nucleation site to adjacent DNA sequence.

    A. the activity of adjacent promoters
    B. the molar concentration of coating proteins
    C. activity of a methylase
    D. activity of a deacetylase
    E. A and B

8. Considering that dosage compensation and the formation of facultative heterochromatin ensures an equimolar amount of product from each chromosome within a species, a female of a particular species maintains a variegated coat color. The gene for coat color is carried on the X chromosome; what is the most likely explanation for this observation?

    A. the female is heterozygous for coat color and is expressing both alleles in each cell
    B. there is no logical explanation for the observation
    C. the female maintains both X chromosomes each maintaining silenced domains
    D. the female is a mosaic of cells each expressing one X chromosome
    E. B and C

9. Considering that patterns of gene repression are passed from primary cells to daughter cells, these *Drosophila* proteins, which maintain chromo domains, can bind inactive chromatin in primary cells and perpetuate the inactivity in resulting daughter cells even in the absence of the sequence specific repressor proteins.

    A. the trithorax protein group (trxG)
    B. the SIR3/SIR4 protein group
    C. the polycomb protein group (Pc-G)
    D. the p300/CBP protein group
    E. C and D

10. The examination of prion related diseases and hence, epigenetic inheritance, in mice reveals that when a mouse PrP-Sc protein is placed into a recipient mouse, the mouse dies, however the hamster PrP-Sc protein has no effect on a recipient mouse. How can this be explained?

    A. the hamster PrP-Sc protein does not normally cause disease even in hamsters
    B. the PrP-Sc and PrP-c proteins must have matched sequences to cause disease
    C. the hamster PrP-Sc protein and mouse PrP-c protein have different sequences
    D. the hamster PrP-Sc and mouse PrP-c proteins maintain different covalent additions
    E. B and C

Answers:

A. Recall and short answer

1. Yes, a transcription factor such as an activator must bind sequence specifically and recruit chromatin-remodeling complexes to the chromatin resulting in promoter activation.

2. DNAase I susceptibility may be a consequence of the movement of the nucleosome core along the DNA or complete removal of the histone core particle by chromatin remodeling enzymes. The disruption of the nucleosome-DNA association provides DNAase I access to the nucleotide sequence resulting in DNA cleavage by enzyme.

3. The remodeling complexes use energy from ATP hydrolysis to twist DNA on the core histone surface. This twisting facilitates the disruption of the weak interactions between the DNA double helix and the histone core resulting in movement of the DNA.

4. The binding of chromatin remodeling complexes such as NURF requires binding by activator proteins. The activator GAGA facilitates NURF binding resulting in the remodeling of chromatin structure and the creation of new boundary regions.

5. Nucleosomes display some steroid response elements in a fashion that is required for hormone receptor binding. Remodeling of the chromatin displaying these sequences would inhibit activator binding. The nucleotide sequence elements are presented in a rotational phase that facilitates protein-protein interaction between the initiation complex components.

6. acetylation; methylation

7. The phosphorylation of histone H3 facilitates the production of extended euchromatin.

8. histone deacetylases, activation

9. The coactivator CBP can facilitate transcription initiation complex formation by physically linking various activators to the basal apparatus or by acetylating the N-terminal tail of histone H4, thereby disrupting nucleosome structure.

10. The relationship suggests that methylated DNA recruits histone methyltransferase enzymes resulting in the methylation of histone proteins; rendering the DNA within the methylated complex transcriptionally silent.

11. The HP-1 protein triggers the production of heterochromatin in mammals by recognizing and binding the methylated N-terminal tail of histone H3. The extent of heterochromatin propagation is limited by an unknown mechanism but it appears the molar concentration of HP-1 may be a factor that limits the extent of heterochromatin propagation.

12. Humans normally maintain only one transcriptionally active X chromosome. However zygotic females maintain two X chromosomes therefore one X chromosome is converted into facultative heterochromatin. The transcriptional inactivation of one of the X chromosomes reduces the products derived from the X chromosome to reflect the norm for the species. Dosage compensation that is reflected in the inactivation of the X chromosome cannot take place in the

absence of DNA methylation. The typical covalent modifications of inactive DNA including the deacetylation of histone H4 and methylation of CpG sequences are evident on the inactive X chromosome.

13. The CpG islands found within heterochromatin are typically methylated. Therefore methylation of DNA is generally associated with the repression of transcription. There is an inverse relationship between hypersensitive sites and the methylation of CpG islands throughout the genome. Methylation of DNA facilitates the reduction of transcription initiation while hypersensitive sites stimulate the activity of promoter regions thereby activating gene expression.

14. Maintenance methylase; acetylation

15. DNA that is actively expressed is associated with nucleosomes that are acetylated on histone H3 and H4 and to a lesser degree phosphorylated on histone H3. While genes that are poorly expressed maintain methyl addition on CpG islands within their promoter regions and are associated with nucleosomes that maintain methyl additions on 9Lys of histone H3.

B. Analytical and critical thinking

1. (a) The repressor complex recruits proteins that maintain common heterochromatin binding motifs including SET and chromo domains. The SET domain participates in the methylation of 9Lys of histone H3 facilitating further association with proteins including those maintaining the chromo domains. The mammalian proteins maintaining chromo domains include the HP-1 family of proteins and the Pc group of proteins. It is likely that the HP proteins interact by use of chromo domains with sites of methylated chromatin resulting in the silencing of gene expression within the methylated DNA. The Pc family of proteins likely coats the regions of DNA that are coated with HP-1 resulting in the further suppression of transcriptional activity.

(b) The formation of transcriptionally inactive chromatin is dependent upon methylation of 9Lys of histone H3. The elimination of proteins from the repressor complex that maintain SET domains will result in the disruption of repressor function, because the proteins maintaining SET domains participate in the methylation event.

2. The FACT complex chaperone facilitates the dissociation of the H2A-H2B dimer from the histone core and replaces the dimer while maintaining rotational and translational positioning of DNA around the histone core. The FACT activity maintains epigenetic information by conserving rotational and translational positioning of the DNA within the nucleosome.

3. Chromatin remodeling is controlled by cellular signals, which relate changing environmental stimuli to the nucleus resulting in a genetic response. Phosphorylated inositol is a common signal transduction molecule involved in receptor mediated alteration of gene expression. Chromatin remodeling and the resulting activation of gene expression is modulated by inositol tetrakisphosphate (IP4).

4. The repressor complex REST/NRSF recruits the HDAC1-2 proteins that act to deacetylate the N-terminal tails of histone H3 and H4. The deacetylation of histone core proteins is a characteristic of nucleosomes found in heterochromatin. Recruitment of and association with the HP-1 family of proteins follow deacetylation of histone core proteins by HDAC1-2. The HP-1 family of proteins nucleates and propagates heterochromatin formation in mammals resulting in

the elimination of gene expression. The complex will facilitate heterochromatin formation as a result of HP-1 binding however long term inheritable gene silencing will result if proteins including the Pc family of proteins are recruited to the complex.

5. The methylation of DNA and histone proteins is a feature of inactive chromatin. Cancer cells are characterized by abnormal gene expression resulting in uncontrolled cell growth and division. The 90% reduction of methyltransferase action in the transgenic mice results in a severe restriction on the animal's ability to inactivate chromatin via the methylation of DNA or histone proteins. A more severe disease phenotype would be expected if the animal completely lacked methyltransferase action. The animal would have no ability to silence gene expression through the formation of heterochromatin.

C. Extensions and applications

1. Imprinted DNA maintains methyl additions and exists in a transcriptionally repressed state. The term generally describes the difference in behavior between two different alleles inherited from each parent. The allele from one parent is transcriptionally silenced while the allele from the other parent is fully expressed. The transcriptionally repressed state is inherited from generation to generation assuming that the enzyme responsible for the maintenance of the methylated state of the DNA is available following DNA replication. The loss of DNA methylation results in the expression of the second allele encoding insulin-like growth factor II. The dosage of this gene product facilitates accelerated cell development resulting in cancer.

2. Histone and DNA methylation are characteristics of heterochromatin. The expression of gene segments contained within the centromere is the result of the loss of methylation of 9Lys histone H3 within the chromatin of the centromere. The loss of methylation correlates with the loss of argonaute function, which is a demonstrated DNA and histone methyltransferase. The control of gene silencing by RNA is typical of mammalian X inactivation, where the Xist RNA coats the inactive X chromosome inhibiting gene expression. It is likely that the Xist RNA recruits enzymes to methylate DNA and histone proteins furthering the inactivation of the X chromosome.

D. Standardized test format

1.	E
2.	C
3.	B
4.	E
5.	B
6.	C
7.	B
8.	D
9.	C
10.	E

Chapter 24

RNA splicing and processing

General Concepts/processes	Section	Related Questions
Pre-mRNAs are spliced	24.1	A1
Mechanisms of splicing	24.2-24.12	A2,A3,A6,A7,A9,B1,B3,C2,C3,D8,D9,D10
Alternative splicing	24.12	C1,C4
Trans-splicing	24.143-24.14	A8
tRNA processing	24.14-24.17	A10,B2,D1,D2,D3
Pol I and Pol III termination	24.18	D4
mRNA 3' end	24.19	A4,A5
Histone mRNA processing	24.20	D5
rRNA processing	24.21-24.22	D6,D7

Questions:

A. Recall and short answer

1. Define the term "heterogeneous nuclear RNA (hnRNA)."

2. Define the term "transesterification."

3. What is the mechanistic difference between the way that group I and group II introns are spliced?

4. Compare and contrast the way that the 3' ends of mRNA transcripts are generated in prokaryotes compared with eukaryotes.

5. Describe the proteins involved in processing the 3'-end of eukaryotic non-histone mRNA. Include the identity of the proteins and their function in your answer.

6. The mechanism by which 5' and 3' intron splice sites are recognized by the splicing apparatus is often described as a paradox. Why?

7. Describe with words or a diagram the mechanism of intron splicing via the lariat structure.

8. Define the term "*trans*-splicing."

9. Does tRNA splicing follow the GT-AG rule and transesterification reactions? Explain your answer.

227

10. How does the reaction catalyzed by the endonuclease that processes eukaryotic tRNAs differ from endonucleases that cleave DNA?

B. Analytical and critical thinking

1. The GT-AG rule is used to predict intron donor and acceptor splice sites. Since these dinucleotides occur very frequently in genes, describe how you would predict the introns of a previously uncharacterized gene.

2. Describe an unusual example of how RNA splicing can serve to regulate gene expression in yeast.

3. U1 snRNA is thought to select the donor splicing junction by base pairing with the 5'-splice site. The following mutations were made and tested in an in vitro splicing reaction. Do the results support the hypothesis? Explain your answer.

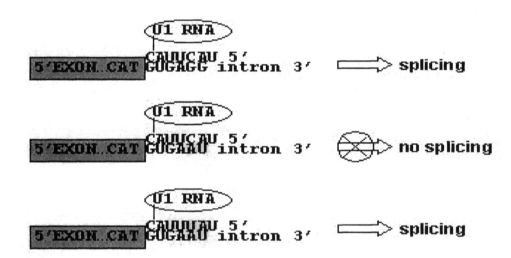

C. Extensions and applications

1. Alternative splicing is known to produce different products from a single gene. First, describe the three types of alternative splicing systems. Then, explain how alternative-splicing mechanisms might occur.

2. The human *Yfg1* gene (your favorite gene) is composed of 4 exons that are normally spliced together to produce a 461 codon open-reading frame. Suppose that a point mutant was identified in the *Yfg1* gene that results in processing of the mRNA without the inclusion of exon 2 . Propose a hypothesis to explain how a point mutation could result in such a mutant mRNA.

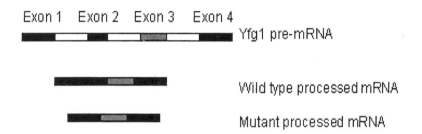

3. Referring to the problem in C2, the length of the second exon is 92 nucleotides. What effect would you predict the point mutation would have on the Yfg1 protein product?

4. Suppose that two processed forms of mRNA are produced from a single pre-mRNA as depicted in the diagram below. Describe a mechanism that could account for the different processed mRNAs.

Pre-mRNA

Processed mRNA variant 1

Processed mRNA variant 2

D. Standardized test format

1. Which of the following is the correct reaction catalyzed by the endonuclease that initiates pre-tRNA splicing reactions?

2. What physical evidence remains at the splice junction of eukaryotic tRNAs after splicing has occurred?

    A. none
    B. They carry a 2'-hydroxyl group on the nucleotides at the splice junction.
    C. They carry a 5'-phosphate at the terminus of the tRNA.
    D. They carry a 2'-3'-cyclic phosphate on one of the nucleotides at the splice junction.
    E. They carry a 2'-phosphate on the nucleotide at the 5' side of the splice junction.

3. Which of the following statements best describes the functions of yeast Ire1p?

    A. It is an integral membrane protein with amino terminal sensor domain in the ER lumen and the carboxyl terminal mRNA splicing domain in the nucleus.
    B. It is an integral membrane protein with amino terminal mRNA splicing domain in the ER lumen and the carboxyl terminal sensor domain in the nucleus.
    C. It is an integral membrane protein with amino terminal mRNA splicing domain in the nucleus and the carboxyl terminal sensor domain in the cytoplasm.
    D. It is a soluble protein with a nuclear localization sequence that serves to move the protein between ER lumen and nucleus.
    E. It is a soluble nuclear protein that becomes activate by Iip1p when malfolded proteins accumulate in the ER lumen.

4. Which of the eukaryotic nuclear RNA polymerases terminate transcription similar to prokaryotic RNA polymerase?

    A. RNA polymerase I
    B. RNA polymerase II
    C. RNA polymerase III
    D. only RNA polymerase I and III
    E. all of them do

5. The mRNAs for histones are unique in that they lack a polyA tail. Which of the following statements concerning how the 3' end of the mRNAs is generated is false?

    A. The 3' terminus of histone mRNAs has a highly conserved stem-loop structure where cleavage occurs.
    B. A stem-loop binding protein is necessary for the cleavage reaction.
    C. The U7 snRNP cleaves 4 to 5 bases downstream of the stem-loop.
    D. The U7 snRNP base pairs with a purine-rich sequence called the histone downstream element (HDE)
    E. The existence of the stem-loop is more important in determining the cleavage site than is its sequence.

6. Which of the following statements about the manner in which the major rRNA's are generated is false?

    A. The 23s, 16s, and 5s RNAs and two tRNAs are transcribed as a single RNA.
    B. Alternative splicing reactions release the individual RNA species.
    C. Exonucleases are involved in rRNA processing.
    D. 3'-5'-exonucleases are involved in rRNA processing.
    E. The exosome, a system involved in mRNA degradation, also plays a role in rRNA processing.

7. Which of the following is not an activity of snoRNAs?

    A. generation of the 3' end of rRNAs
    B. methylation of rRNAs
    C. synthesis of pseudouridine
    D. mRNA degradation
    E. cleavage of the C-N bond of uridine

8. Which of the following intron classes form an intermediate lariat structure?

    A. nuclear RNA
    B. group I
    C. group II
    D. nuclear and group I
    E. nuclear and group 2

9. Which of the following statements about splicing is false?

    A. Introns are sequentially removed from the leftmost intron to the rightmost intron.
    B. Intron removal is not random; there is a preferred order.
    C. Splicing is either autonomous or requires a spliceosome.
    D. The spliceosome is a combination of RNA and proteins.
    E. The spliceosome is assembled on the pre-mRNA.

10. Which of the following is not involved in mRNA splicing?

    A. U3
    B. U4
    C. U5
    D. U6
    E. U7

Answers:

A. Recall and short answer

1. hnRNA is the mixture of RNAs including pre-mRNAs and other transcripts found in the nucleus.

2. Transesterification refers to the breakage of a phosphodiester bond and reformation at a different location.

3. Group I introns use a G nucleotide from solution to carry out the first reaction of transesterification. Group II introns use an A nucleotide within the intron, referred to as a branch site, for the first reaction of transesterification.

4. Transcription termination in prokaryotes depends on specific elements that signal RNA polymerase to end transcription, so transcripts have a clearly defined 3'-end immediately after transcription terminates. By contrast, in eukaryotes there are no specific termination signals and RNA polymerase II transcribes a long 3'-end. After transcription a specific endoribonuclease cleaves the 3'-end of the transcript and then polyadenylation occurs.

5. A complex of proteins processes the 3'-end of transcripts. The complex includes CPSF, which recognizes and binds to the polyadenylation signal (AAUAAA), a factor that stimulates CPSF binding called CstF, which binds to a GU-rich region downstream of the polyadenylation signal, an endonuclease composed of two subunits CFI and CFII, and polyA polymerase PAP, and finally, PABP binds to the polyA stretch added by PAP.

6. The 5' and 3' splice sites are each defined by two nucleotides, GU at the 5'-end and AG at the 3' end. These sequences occur frequently in a gene, yet, mRNA splicing is a very specific process, with only specific GU-AG pairs being recognized by the splicing apparatus. The basis for accurate recognition splice sites is not known.

7. The 2'-OH of the branch-site A reacts with the phosphodiester bond linking the G at the 5'-end of the intron to the exon. The phosphodiester bond is broken and then reformed between the 5'-G and branch-site A. The 3'-OH of the exon, produced by hydrolysis of the initial cleavage reaction, reacts with the phosphodiester bond linking the G at the 3'-end of the intron with the exon. The bond is broken and a new bond is formed linking the two exons flanking the intron. The lariat-shaped intron is released.

8. *Trans*-splicing refers to an intermolecular splicing reaction between two RNAs.

9. No. Certain tRNAs in yeast have introns. These introns do not follow the GT-AG rule. tRNA splicing depends on the recognition of an RNA secondary structure by a multisubunit endonuclease called Sen34/Sen2. The exons are joined by another enzyme called RNA ligase in an ATP-dependent reaction.

10. The splicing endonuclease produces a 5'-hydroxyl and a 2'-3'-cyclic phosphate at the cleavage site, whereas a DNA endonuclease generates a 5'-phosphate and 3'-hydroxyl at the cleavage site. The RNA endonuclease recognizes a RNA secondary structure, and the DNA endonuclease recognizes a DNA sequence, in the case of a site-specific endonuclease, or any DNA sequence in the case of a general endonuclease.

B. Analytical and critical thinking

1. There are several approaches that can be used. First, is the sequence of a cDNA derived from the gene available? If so, it can be aligned with the gene sequence to delineate the boundaries of the introns. The second approach would be to identify open reading frames within the gene sequence. Exons will have extended open reading frames, whereas introns will contain stop codons. The border areas between open reading frames and stop codons can be scanned for GT-AG pairs. Finally, the correct GT-AG pair can be identified if removal of the intron links the flanking exons into a continuous open reading frame. A third criterion would be to identify a branch-site sequence between the GT and AG borders of the intron between 18 and 40 nucleotides upstream of the 3' splice site. However, since branch-site sequences are not well conserved, this criterion is most ambiguous.

2. *GENES VIII* describes the RNA splicing event that regulates Hac1-controlled chaperonin expression in response to unfolded proteins in the endoplasmic reticulum. Unfolded proteins in the ER are detected by Ire1, which is an RNA endonuclease that splices away a specific section of the Hac1 mRNA responsible for Hac1-mRNA degradation. The stabilized Hac1 mRNA is then translated and goes on to regulate expression of chaperonins.

3. The hypothesis is supported, but not proven. Mutations in the 5' donor site that abolish splicing are suppressed by mutations in U1 RNA. Since the number of normal base pairs in the mutant is not changed it is possible that the pattern of pairing is important. Removing the G:C pair near the 5' end of U1 might need two weaker A:U pairs to compensate.

C. Extensions and applications

1. Any example is acceptable, even a hypothetical one. However, *GENES VIII* describes the SV40 T/t antigens and E1A systems in which two or more different 5' sites are spliced to a single 3' site, the *D. melanogaster tra*, *dsx*, tropomyosin, and P-element systems in which exons are skipped or substituted by use of alternative 3' sites, alternative 5' sites, and alternative 5' and 3' sites. One way to produce splice variants is through existence of proteins that bind to RNA sequences and block the use of splice sites or promote the use of splice sites. One example is the SF2/SF5 system that controls the T/t antigen and E1A systems. SF2 blocks the use of the 5' splice site most distant from the 3' splice site. SF5 counteracts the effect by blocking use of the 5' splice site closest to the 3' splice site.

2. The omission of exon 2 in the mutant could have resulted from a single nucleotide change that altered splicing. For example, a mutation in the acceptor splice site at the 3' end of intron 1 could result in exon 2 being skipped over.

3. The omission of exon 2 results in fusion of exon 1 with exon 3. Since exon 2 is 92 nucleotides the reading frame would be lost and it is likely that a short truncated protein would be produced.

4. The alternative use of 3' splice sites could be driven by a factor that binds to the normal splice site at the 3'-end of intron 3. When this site is blocked, the splicing apparatus could skip to the next 3' splice site. What looks like intron 4 would have to have a defective 5' splice site so that

exon 4 would never splice to 5 and 6. Also intron 4 would need a regulated polyA site so that exon 5 and 6 and intron 5 would not be present in the mRNA as shown.

D. Standardized test format

1. B
2. E
3. A
4. D
5. C
6. B
7. D
8. E
9. A
10. E

Chapter 25

Catalytic RNA

General Concepts	Section	Related Questions
RNA splicing and editing	Introduction	B1, C2, D2
Group I introns (self-splicing)	25.2, 25.3	A3, A4, A5, A6, B6, D2, D3
Ribozymes	25.4	A1, D4, D11
Group I introns and mobility	25.5	A3, A4 A5, A6, B3, C2
Group II introns	25.6	A7, B6, D5, D6
RNAase P and viroids	25.7, 25.8	A8, A10, D7
Mechanisms of RNA editing	25.9, 25.10	B4, B5, C1, D10
Processes/structures		
RNA editing	Introduction	A2, A12, B4, C1, D8, D10
IGS (Internal Guide Sequence)	25.4	B2
Viroids and Virusoids	25.8	A9, A10, D7
Protein splicing	25.11	A11, A12, C2, D9
Examples, Applications and Techniques		
RNAase P	25.7	A8, D1
Apolipoprotein-B and glutamate	25.9	

Questions:

A. Recall and short answer

1. Catalytic properties are known to exist in RNA. A _____ is the general term used to describe an RNA with catalytic activity.

2. Like RNA splicing, _____ is the process which involves the insertion of bases (usually uridine) in mitochondrial products of mitochondrial genes and in certain "lower" eukaryotes.

3. Self-splicing is a property of transcripts of the rRNA genes in *Tetrahymena* and *Physarum*. It is a property of _____ introns.

4. Contrary to the requirements of alternative splicing of pre-mRNAs, group I splicing has minimal chemical requirements. What are these requirements?

5. What evidence indicates that group I intron self-splicing has no energy requirements?

6. Self-splicing of group I introns is dependent on the development of a complex secondary structure. Such a structure is equivalent to the generation of active sites in conventional

_____.

7. Group II introns can code for a _____ that generates a cDNA.

8. Ribonuclease P is an *E. coli* ribonucleoprotein that involves tRNA processing through endonuclease activity. Originally it was thought that the protein provided the catalytic function, however it turns out that both RNA and protein components are involved in RNAase P function *in vivo*. What experimental approach was instrumental in supporting this conclusion?

9. Small plant RNAs that are capable of self-cleavage are grouped into two classes. Name these two classes.

10. What observation indicates that the self-splicing property of viroids and virusoids is dependent on a structural conformation?

11. Mutation and DNA restructuring through crossing over can provide for genetic variability at the level of DNA. The process by which information changes at the level of RNA is called_____.

12. Protein splicing has the same effect as RNA splicing in that a sequence that is eventually represented in the final protein product is missing from the initial product. The terms which apply to RNA splicing (exons and introns) have been adapted to protein splicing. What are these analogous terms?

13. How are inteins recognized?

B. Analytical and critical thinking

1. Present an assay system and/or experimental protocol that indicates that the *Tetrahymena* intron is capable of "self-splicing" in *E. coli*.

2. Describe the involvement of internal guide sequences (IGS) as a ribozyme behaves as a nucleotidyl transferase.

3. Compare and contrast the mobility characteristics of group I and group II introns.

4. RNA editing is known to involve both *cis*-acting and *trans*-acting processes. Explain what each of these processes is in terms of RNA editing.

5. Provide a brief explanation of RNA editing and protein splicing. Present a possible rationale for the evolution of these two processes.

6. Some group I and group II mitochondrial introns have open reading frames. What functions can be ascribed to such sequences?

C. Extensions and applications

1. RNA editing modifies genetic information at the transcript level by primarily involving cytidine (C) to uridine (U) conversions and the reverse. In a study of the hornwort, *Anthoceros formosae*, 49 RNA editing sites were described that involved 509 C-to-U and 433 U-to-C conversions leading to a replacement of 868 out of 24275 (3.6%) amino acids studied (Kugita et al. 2003 *Nucleic Acids Research* 31(9) 2417-2423). Ninety-seven percent of the conversions involved the first or second base of the mRNA codon and lead to substantial changes in coding (altered amino acid specification or nonsense codon shifted to sense codon) even though the general distribution of pyrimidines in the mRNA is approximately 41.4%(first position), 53.0%(second position), and 52.5%(third position). There is a strong bias toward the conversion of nonsense codons to sense codons in protein-coding genes. Given this information, what would you consider the function of RNA editing? Why?

2. Intron homing is the process that, as related to mobility, ensures that a given intron perpetuates itself by insertion into a single target site and nowhere else in the genome. The intron encodes a homing endonuclease that cleaves at a specific target site for intron insertion. Like group I introns, inteins are protein-splicing elements that encode conserved sequence blocks characteristic of homing endonucleases. Derbyshire et al. 1997 (Proc. Natl. Acad. Sci. USA 94:11466-11471) examined the possibility of a common evolutionary origin of mechanisms responsible for mobile introns and inteins. What kinds of observations would support such a common origin?

D. Standardized test format

1. The enzyme ribonuclease P contains a single RNA molecule bound to a protein. The _____ component has the ability to cleave _____ while the protein component is probably used to _____.

      A. RNA; tRNA; maintain the structure of the catalytic RNA.
      B. DNA; tRNA; cleave the DNA double strands.
      C. RNA; mRNA; enzymatically attack the RNA.
      D. RNA; tRNA; cut the RNA at the 3' end.
      E. RNA; rRNA; stabilize the DNA component.

2. Self-splicing occurs by _____ reactions in which bonds are directly exchanged.

      A. ionic
      B. sulfonic
      C. hydronic
      D. calcification
      E. transesterification

3. Group I introns form a secondary structure with 9 duplex regions. The core regions, P3, P4, P6, and P7 have _____ activity.

    A. intein
    B. intron
    C. exon
    D. catalytic
    E. membranous

4. Catalytic reactions of a ribozyme involve transesterification reactions between a group in the substrate-binding site and a group in the _____ binding site.

    A. G
    B. A
    C. T
    D. R
    E. E

5. Some group II introns code for a _____ that can generate a DNA copy of the RNA which transposes by a _____-like mechanism.

    A. polymerase; viral
    B. reverse transcriptase; retroposon
    C. reverse transcriptase; RNA
    D. retroelement; catalase
    E. none of the above

6. Intron homing is the process that involves motility of an intron such that

    A. it can integrate anywhere in the genome.
    B. it can only perpetuate itself in centromeric DNA.
    C. it inserts into a single target site in the genome.
    D. it allows the intron to integrate with the help of RNAase P.
    E. it inserts into any site other than in exons.

7. A _____ is an infectious RNA molecule that functions independently without encapsidation in a protein coat.

    A. virus
    B. capsulon
    C. prion
    D. antiprion
    E. viroid

8. The process of _____ involves the formation of mRNAs which differ from the DNA from which it was transcribed.

    A. RNA surfing
    B. RNA editing
    C. DNA surfing
    D. DNA editing
    E. RNA transfixing

9. Inteins differ from introns in that inteins involve _____ splicing and introns involve _____ splicing.

    A. protein; RNA
    B. noncatalytic; catalytic
    C. autonomous; nonautonomous
    D. RNA; protein
    E. RNA; DNA

10. RNA editing is a complex process that generally involves two distinct enzymatic processes: _____ and _____.

    A. detoxification; condensation
    B. involution; exolution
    C. cleavage; ligation
    D. respiration; oxidation
    E. oxidation; reduction

11. Self-splicing is a property of two groups of introns that are found primarily among lower eukaryotes, prokaryotic systems and mitochondria. The information necessary for self-splicing reactions resides primarily in

    A. the intron sequence but is assisted by proteins *in vivo*.
    B. the nucleus of the host organism.
    C. chromosomal DNA that is not transcribed.
    D. tRNA.
    E. the proteins associated with the cell membrane.

Answers:

A. Recall and short answer

1. ribozyme

2. RNA editing

3. Group I

4. monovalent cation, divalent cation, and a guanine nucleotide cofactor

5. a triphosphate is not needed; GTP, GDP, GMP, and guanosine itself can all be used

6. proteins

7. reverse transcriptase

8. Mutations in either the gene for the RNA component or the gene for the protein component can inactivate RNAase P activity *in vivo*.

9. viroids and virusoids

10. The generation of activity can be altered by heating and cooling. Its active conformation can be generated by denaturation followed by renaturation.

11. RNA editing

12. exteins and inteins

13. Inteins can be recognized by the existence of homologous genes that lack an insertion. They have an N-terminal serine or cysteine and a C-terminal asparagine.

B. Analytical and critical thinking

1. One experimental procedure could involve placing a putative *Tetrahymena* intron in a marker gene which is then inserted into *E. coli* through transformation or transfection. If the marker protein is made, then successful splicing in *E. coli* would be supported. A common assay system involves the luciferase gene or the gene for β-galactosidase. In the latter case, if the intron is engineered into the β-galactosidase gene and the medium is supplemented with X-gal (5-bromo-4-chloro-3-indolyl-β-D-galactopyranoside), a blue colony will occur if self-splicing is successful. This is because the protein can only be translated successfully if the intron is removed.

2. The IGS serves as a positioning device for intermolecular rearrangements by aligning the substrate for the transesterification reaction. The reaction involves a G-OH attack of a substrate in which a C is transferred to a 3'-G. A reverse reaction can occur in which a carbon is added to a polyribonucleotide substrate.

3. Certain introns of both group I and group II contain sequences that can be translated into proteins. These proteins are usually related to mobility which involves insertion into another site. Group I introns migrate by DNA mediated mechanisms whereas group II introns migrate through RNA-mediated mechanisms. Some group I introns can code for endonucleases that make double-stranded breaks in DNA target sites. Some group II introns code for a single protein which contains reverse transcriptase, maturase, and endonuclease activities.

4. RNA editing based on *cis*-acting processes is dependent on components intrinsic to the RNA being processed. We might envision this as the generation of a folded RNA which causes editing. Trans-acting elements would be dependent on editing enzymes distinct from the RNA in which the editing occurs.

5. RNA editing allows for changes to occur in an RNA product while protein splicing is a reaction similar to intron removal during RNA processing but in this case, proteins are involved. Inteins are spliced out of flanking exteins. RNA editing provides an opportunity to correct errors in RNA sequences or to generate protein variety as in the case of apolipoprotein-B and the glutamate receptor. Protein splicing may be related to various forms of nucleolytic attack as evidenced by a possible relationship to homing endonucleases that sponsor group I intron motility. It is likely that protein splicing, like various forms of nucleic acid splicing, allows for error correction and/or protein variety.

6. Proteins coded by groups I introns are endonucleases that cleave target double-stranded DNAs to initiate gene conversion. Proteins coded by group II introns have endonuclease activity that enables an RNA copy of the intron to be inserted into target sites.

C. Extensions and applications

1. Because the distribution of base changes resulting from RNA editing is not random (preferring the first and second coding bases) and there is a strong bias toward correcting the most deleterious of mutations (nonsense), it seems likely that RNA editing evolved to restore function to mutant proteins. Given that the conversions involve pyrimidines, and there is no strong bias of pyrimidines in the first and second codon base positions, the preference for conversions in those positions (first and second bases) is not based on a high pyrimidine frequency. Rather, there must be selective pressure to have RNA editing concentrated on those bases which are most likely to cause significant coding repair.

2. In general, preserved integrity of sequence and structural aspects of genetic elements supports common evolutionary origin. Functional relationships between intron and intein splicing could be tested by deletion or mutation analysis. Abolishment of both functions by a single mutational event suggests common functional origin. Phylogenetic distributions of splicing elements may also shed light on their origins. If such elements share common distribution in diverse phylogenetic groups then common origin is suspected. Independent origin is suspected if intron-splicing and intein-splicing elements are separately distributed among different species.

D. Standardized test format

1. A
2. E
3. D
4. A
5. B
6. C
7. E
8. B
9. A
10. C
11. A

Chapter 26

Immune diversity

General Concepts	Section	Related Questions
Chain structure	26.3-26.6	A3, D1
B-cell receptor	26.11; 26.17	A7, B4,
T-cell receptor	26.18-26.19	A12, D6, D10, D7,
MHC	26.19-26.20	A14, B3, C2, D8
Processes/structures		
Clonal selection	26.2; 26.17	A11, D2,
Recombination	26.4-26.9	A2, A4, A5, D3
Deletion / inversion / mutation	26.14-15	A10, B1,
Class switching	26.11-26.13	A7, A8, A9, B1,
Allelic exclusion	26.10	A6, D4, D5
Adaptive immunity	26.17; 26.21	A1, A13, D9
Innate immunity	26.21	A15, B2, B5, C1

Questions:

A. Recall and short answer

1. Immature B lymphocytes that bind an antigen increase in number by clonal amplification and mount the _____response, while subsequent exposures to the same antigen activates memory cells resulting in _____ response.

2. Considering that the genes encoding the various components of T cell receptors and B cell derived antibodies are inherited and the genetic response to any particular antigen is not inherited by the offspring of the organism, when must the genetic recombination producing the antigen specific lymphocyte take place?

3. An individual antibody maintains 2 identical light chain polypeptides and 2 identical heavy chain polypeptides, each consisting of constant and variable regions. Considering that the genome contains gene segments encoding for approximately 10 constant regions and over 100 variable regions, which segments of the polypeptides must recognize antigens? Why?

4. Considering that the V, D and J categories of gene segments may lie in order along the DNA sequence and that duplication of any category will result in an ineffective immunoglobulin, what characteristic of the nucleotide sequence ensures proper recombination order? How?

5. The genetic recombination of heavy chain gene segments (V, D and J) results in approximately 1/3 aberrant genes containing altered reading frames. How are these nonfunctional genes considered a reasonable trade off for increased antigen binding sensitivity?

244

6. Gene segment recombination during antibody formation can be termed nonproductive. What structural nucleotide configuration reflects nonproductive recombination?

7. Within lymphocytes, class switching refers to the ability of a cell to halt production of one immunoglobulin and begin the secretion of a second. The process is mediated by a change in CH gene segment expressed within the cell. How does this change reflect antigen specificity? Explain.

8. All lymphocytes begin productive life as immature cells synthesizing the immunoglobulin IgM. However, many lymphocytes enter circulation as mature cells expressing different classes of immunoglobulin. How can this observation be explained?

9. Describe the relationship between RNA editing and the mature B lymphocytes ability to secrete antibody.

10. Mutations within the variable chains of immunoglobulins are essential for matching antigen diversity. In addition to the sequence diversity introduced during V-J or V-D-J recombination somatic mutations are crucial for generating changes in the variable chains. Name the enzymes responsible for somatic mutation generation. Do all mutations increase variable chain diversity? Why or why not?

11. Briefly describe why the secondary immune response mediated by B lymphocytes involves no somatic mutation or immunoglobulin type switching.

12. It is speculated that the genetic diversity of the TCR region, generated within T lymphocytes, is generated in much the same way as the genetic diversity is produced in the Ig genes of B cells. Provide evidence to support this speculation.

13. The secondary immune response is critical for the efficient and timely elimination of pathogens. This very rapid immune response is due to what characteristics of the B lymphocytes?

14. The histocompatibility locus in man contains many genes whose proteins function within the immune system. The immune related proteins encoded by these genes include which groups?

15. Organisms that do not maintain an adaptive immune response (acquired immunity) like *Drosophila* depend upon what to fight infection? What factors are responsible for the activation of their resistance pathway?

B. Analytical and critical thinking

1. Intestinal B cells begin life producing IgM and upon maturation and movement into the gut largely produce immunoglobulin A (IgA). Work by Fagarasan et al. (2002 *Science* 298:1424-1427) demonstrated that B cells lacking activation induced cytidine deaminase activity accumulated elevated levels of IgM and failed to produce IgA. Explain the observation.

2. B cell memory is essential for the ability of the acquired immune response to function. However plasma cells, derived from antigen activated B cells, maintain a limited life span of a few months placing a limitation on the lifespan of antigen memory. Bernasconi et al. (2002 *Science* 298:2199-2202) demonstrated that in addition to specific antigen, other "nonspecific" factors maintain B cell memory by driving memory B cell proliferation and differentiation. These "nonspecific" factors act indiscriminately through the TLR4 and TLR 9 receptors to maintain a broad spectrum of antibody specificities within the

serum by acting on all available memory B lymphocytes. What are these "nonspecific" factors and where are they derived? What do these results suggest regarding the interaction between the innate and acquired immune responses?

3. T-cell activation is dependent upon the interaction between the antigen, MHC protein(s) and the TCR. Messaoudi et al. (2002 *Science* 298:1797-1800) demonstrated that mutations within the genes encoding the H-2k gene in mice produced increased diversity and avidity in activated T-cells when exposed to a viral pathogen in comparison to control mice. The T-cell action in the transgenic mice produced a 5x greater survival against specific viral infection. What proteins did the mutated genes encode? Considering the genes involved what do the results infer regarding immune gene polymorphism and the ability of T-cells to recognize antigens and fight disease?

4. Recent work by Wardemann et al. (2003 *Science* 301:1374-1377) demonstrated that 55 to 75% of all antibodies expressed by human immature B cells maintained self-reactivity and of those 55.2% were polyreactive to double stranded DNA, insulin and lipopolysaccharide. However, these immature B-cells are eliminated during development at two checkpoints. What physiological drawback would be a result of high levels of self-reactive mature B-lymphocytes in circulation? Is the loss of at least half of the immature B-lymphocytes the cost of increased immune diversity? Why?

5. A common feature of plants and animals is the capacity to recognize potentially pathogenic microorganisms and mount an effective immune response. Since the immune responses are shared it is likely that the pathways responsible for the activation of the immune response are common in plants and animals. Such conserved elements have been identified including the Toll like receptors (TLRs), which function to induce the immune response in plants and animals. Recent work by Yamamoto et al. (2003 *Science* 301:640-643) demonstrated that the pathway for TLR3 action in mammals includes adapter proteins, protein kinases, MAP kinase and transcription factors. What immune response is elicited by activation of TLRs and what antigenic factors are likely to induce such a response? What do these results suggest regarding pathway organization in plants?

C. *Extensions and applications*

1. The innate immune system is highly conserved in nature. The response pathway is activated by ligand binding of the TRL followed by recruitment and activation of unique protein kinases subsequently leading to transcription factor activation. Recent work by Picard et al. (2003 *Science* 299:2076-2079) indicates that genetic based deficiencies in TRL activated protein kinases in humans led to opportunistic infection by only two bacteria. How many TLRs are thought to activate innate immunity in humans? In terms of fighting potentially pathogenic bacteria, what is the division of responsibility for innate and adaptive immunity in humans? If the same kinase deficiencies were observed in *Drosophila*, what would be the expected outcome? Why?

2. The major histocompatibility locus (HLA) encodes amongst other proteins, those proteins involved in human T-cell immunity. Antigen specific T-cell immunity is encoded by the HLA locus and is therefore HLA-restricted. Moore et al. (2002 *Science* 296:1439-1442) demonstrated that mutations in the Human Immuno Virus (HIV-1) facilitated the avoidance of HLA restricted immune responses. HIV-1 genetic polymorphisms are generally associated with HLA class I alleles, therefore establishinga that HIV-1 was unrecognized by T-cell mediated immunity. What aspects of T-cell mediated immunity do proteins encoded by HLA class I alleles participate? Does the limited genetic nature of the class I MHC provide a selective pressure for the evolution of viruses to defeat T-cell mediated immunity? Why or why not?

Does the same potential for avoidance of the immune response due to antigen mutation exist in B-cell mediated immune responses? Why or why not?

D. *Standardized test format*

1. The genetic diversity observed in the production of antibodies within B lymphocytes is a result of

> A. somatic recombination.
> B. somatic exclusion.
> C. cytidine deaminase action.
> D. clonal selection.
> E. A and C

2. Clonal selection following the primary immune response results in the production of B lymphocytes that

> A. produce immunoglobulins recognizing one antigen.
> B. produce immunoglobulins recognizing many antigens.
> C. lack the light chain assembly.
> D. undergo continued somatic mutation generating new V gene sequences.
> E. A and D

3. The relationship between the RAG proteins and the genetic recombination of immune genes is that

> A. recombination is inhibited in the presence of RAG 1.
> B. the RAG proteins catalyze genetic recombination in bacteria.
> C. the RAG proteins are responsible for sequence recognition and DNA cleavage.
> D. the RAG proteins disrupt recombination.
> E. C and D

4. A B lymphocyte that has undergone a productive rearrangement normally maintains these characteristics

> A. production of a single type of light and heavy polypeptide chain.
> B. continued somatic recombination of immune genes.
> C. continued somatic mutation catalyzed by cytidine deaminase.
> D. silencing of immune gene alleles on one of the two chromosomes.
> E. A and D

5. Considering that the production of immunoglobulins involves an ordered series of events, of the actions listed below this action must occur first.

> A. antigen binding.
> B. somatic recombination.
> C. allelic exclusion.
> D. clonal expansion.
> E. heavy chain assembly.

6. These cell surface proteins present foreign antigens to T helper cells, effectively inducing a cellular response.

        A. MHC class II proteins.
        B. MHC class I proteins.
        C. complement proteins.
        D. CD3 proteins.
        E. A and B

7. Immune diversity in T lymphocytes is generated by which of the following mechanisms.

        A. somatic recombination of immune genes
        B. somatic mutation
        C. activity of cytidine deaminase
        D. α-chain rearrangement
        E. A and D

8. These characteristics are shared by the MHC class of proteins.

        A. all function as dimers
        B. all composed of different subunits
        C. require phosphorylation for function
        D. bind antigen cause cell lysis
        E. induce clonal expansion of B-cells

9. The effectiveness of the innate immunity system is derived from

        A. somatic recombination of immune genes.
        B. recognition of predefined patterns typical of pathogens.
        C. interaction with MHC class II proteins.
        D. ability to generate memory cells.
        E. A and D

10. The $\alpha\beta$TCR of the T lymphocyte is said to have the "dual" capacity for recognizing what?

        A. macrophage and T helper cell
        B. H2 locus and complement
        C. antigen and MHC protein
        D. lipopolysaccharides and endotoxins
        E. A and D

Answers:

A. Recall and short answer

1. primary immune; secondary immune

2. The genetic recombination producing antigen specific lymphocytes takes place during the lifetime of the organism. The production of antigen specific lymphocytes is continual for the lifetime of the organism.

3. The variable regions of the immunoglobulins maintain highly variable specificities resulting in the ability to bind and respond to diverse antigens.

4. The recombination between the V, D and J gene segments is facilitated by two consensus sequences that maintain different spacing. The recombination follows the simple rule that segments maintaining one type of spacing must join with a gene segment maintaining a second type of spacing. Therefore like gene segments, maintaining the same spacing around the consensus sequence, cannot join eliminating ineffective recombination events between the V, D and J segments.

5. The insertion or deletion of nucleotides at joining regions potentially alters reading frame of the resulting gene. If a gene segment is combined out of reading frame subsequent translation of the mRNA generated from the gene terminates prematurely and a nonfunctional protein is generated. However, the manipulation of joint sequence potentially increases diversity within the gene sequence resulting in maximum diversity in the resulting immunoglobulin. Therefore, the cost of nonfunctional production is outweighed by the increased diversity observed in immunoglobulins.

6. A nonproductive recombination of gene segments encoding the V, D and J segments appears as an incomplete recombination in which the gene maintains only two of the three segments or is in some fashion producing a nonfunctional immunoglobulin protein.

7. The variable region of the immunoglobulin is not changed within the process of class switching therefore the antigen specificity is not altered.

8. All lymphocytes have the ability to switch the class of immunoglobulin that they secrete. The process involves a substitution in the type of CH region that is expressed as part of the immunoglobulin.

9. The exposure to and recognition of antigen induces the cellular differentiation of B-lymphocytes resulting in a change in RNA editing. The altered RNA editing facilitates the conversion of the membrane bound IgM to a secreted version of the immunoglobulin.

10. Cytidine deaminase and uracil glycosylase; no, like all mutations to DNA only those that alter the amino acid that is encoded by the mutated gene sequence will increase the diversity of the chain sequence.

11. The B-lymphocytes that are activated within the secondary immune response have been previously selected by antigen in an earlier primary immune response. Therefore immunoglobulin structure that was previously selected by the invading pathogen is maintained. This response is characterized by clonal expansion of the B-lymphocytes producing the required antibody. Genetic changes due to somatic mutation or type switching will be counterproductive to the ability of the immunoglobulin to recognize the antigen.

12. The organization of the genes encoding the TCR proteins and Ig proteins is similar. In addition, the recombination of V, D, J and C segments is similar to the recombination that is observed in the generation of immunoglobulins. The recombination of gene segments is directed by the spacing of consensus sequence in both systems. Similarly, the introduction of diverse joint sequence between the recombined gene segments occurs in both systems however somatic mutation does not occur at the TCR gene loci.

13. B cell memory

14.   MHC class I
        MCH class II
        complement

15. The innate immune system; The innate immune system is triggered by pathogen-associated molecular patterns (PAMPS) which comprise a group of bacterial and fungal proteins that are broadly distributed across a wide range of organisms. This group of molecules includes, amongst others, the flagellin protein, lipopolysaccharides and peptidoglycans.

B. Analytical and critical thinking

1. The class switching of immunoglobulins is facilitated by the action of cytidine deaminase. Therefore B cells lacking cytidine deaminase would lack the ability to switch classes of immunoglobulin.

2. The nonspecific factors that function through the TLR4 and TLR9 receptors are pathogen associated molecular pattern molecules (PAMPs), which are generated from bacterial or fungal invaders. The TLR receptors are Toll like receptors that function within the innate immune system. The results of the study imply that the molecules activating the innate immune system also activate the plasma B cells inducing a secondary immune response resulting in the production of diverse antibodies maintaining broad antigen specificity.

3. The mutated genes encoded the MHC class I proteins; the results of the experiment imply that the diversity of the MHC class I proteins directly influence the ability of the T-cell to recognize antigen. The recognition of antigen/MHC protein complex by the T-cell results in the activation of the T-cell. Therefore, increasing the polymorphic nature of the genes encoding the MHC class I proteins likely increases the antigenicity of the displayed antigen.

4. Self reactive antibodies recognize proteins that reside on the surface of self-cells as antigens. This process targets these cells for destruction by the adaptive immune system. The physiological ramifications would include autoimmune diseases specific to the cell(s) that are targeted as antigenic. Yes, the production and destruction of self-reactive immature B cells is the cost of increased immune diversity. Diversity is generated by genetic recombination, nucleotide insertion and enzyme catalyzed mutation. These random mechanisms generate a tremendous level of immune diversity resulting in immature B-cells specific for diverse antigens. A consequence of the random generation of immune diversity is the generation of immature B-cells that display self-reactivity.

5. The innate immune system is activated by TLRs (Toll like receptors); The antigenic factors that are recognized by the TLRs include nonspecific molecules that contain pathogen associated molecular patterns (PAMPs). Molecules in this category include bacterial flagellin, bacterial lipopolysaccharide and bacterial peptidoglycan. Considering that the receptors and antigens are similar in plants and animals it is

likely that plants maintain adapter proteins, protein kinases, MAP kinases and transcription factors that are responsible for the innate immune response.

C. Extensions and applications

1. (a) Humans contain ~ 20 Toll like receptor genes; however, it appears only two bacteria are targeted specifically by the innate immune system. The results suggest that the adaptive immune system is responsible for mounting an immune reaction to the vast majority of potentially pathogenic organisms. However, the innate immune response is likely generated in addition to the adaptive immune response in many circumstances. The results of previous studies imply that the molecules activating the innate immune system also activate the plasma B cells inducing a secondary immune response resulting in the production of diverse antibodies maintaining broad antigen specificity.

(b) If a kinase deficiency rendered the innate immune system dysfunctional in *Drosophila* a severe phenotype would result. *Drosophila* maintain no adaptive immune system therefore the kinase deficiency would render them defenseless against potentially pathogenic organisms.

2. (a) The human HLA class I allele encodes the MHC class I proteins. It appears likely that the limited genetic nature of the HLA class I allele results in selective pressure for the development of resistance by various pathogens. Recall that Messaoudi et al. (2002 *Science* 298:1797-1800) demonstrated that mutations within the genes encoding the H-2k gene (MHC class I proteins) in mice produced increased diversity and avidity in activated T-cells when exposed to a viral pathogen in comparison to control mice. Therefore it suggests that the limited genetic diversity of the HLA locus can be taken advantage of by highly mutable viruses.

(b) The avoidance of B-cell mediated immunity by a potentially pathogenic organism is unlikely due to the tremendous immune diversity generated within the process of immunoglobulin development.

D. Standardized test format

1.   E
2.   A
3.   C
4.   E
5.   B
6.   A
7.   B
8.   A
9.   B
10.   C

Chapter 27

Protein trafficking

General Concepts	Section	Related Questions
Protein glycosylation	27.2, 27.3	A1, A2, B1, B4, D1, D2
Vesicular transport of proteins	27.4-27.6	A3-A5, B3, C2, D3, D4
Vesicle budding and fusion	27.7-27.10	A6-A7, B2, C1, D5-D7
Protein targeting and vesicle sorting	27.11, 27.12, 27.14	A8, B5, D8
Endocytosis	27.5, 27.15, 27.16	A9, A10, D9, D10

Questions:

A. Recall and short answer

1. What is the function of dolichol in the ER membrane?

2. What is Endo H, and why is it a useful tool for studying glycoprotein secretion?

3. Which Golgi stack receives vesicles from the ER?

4. What are two very important functions of coat proteins?

5. COP-I-coated vesicles are primarily involved in (anterograde/retrograde) movement of proteins between Golgi cisternae. COP-II-coated vesicles are involved in (anterograde/retrograde) transport between the ER and the Golgi.

6. What do ARF and dynamin have in common? What functions do they mediate during vesicle formation?

7. Binding and fusion of vesicles with their target membrane requires interactions between two different sets of vesicle-target protein complexes. First, the _____ _____ (two words) on the target membrane binds to a _____ protein in the coatomer of the vesicle. This results in the initial anchoring of the vesicle to the target membrane. Fusion of the two membranes is mediated by the interaction of a _____ protein on the vesicle membrane with a _____ protein on the target membrane.

8. How do lysosomal proteins get packaged into the appropriate vesicles?

9. What is the difference between early and late endosomes?

10. The LDL receptor-LDL complex is internalized via receptor-mediated endocytosis. What happens to the LDL particle? What happens to the LDL receptor?

B. Analytical and critical thinking

1. Discuss two differences between the processes of N-linked and O-linked glycosylation.

2. Protein coats are required for vesicle budding, but inhibit vesicle fusion. Explain.

3. Briefly contrast the two models for anterograde transport through the Golgi.

4. Lys1 is a group II single-pass integral membrane protein (see section 8.14 for review) found in the membrane of lysosomes. A biochemical test reveals that the Lys1 protein is highly glycosylated. Do you predict that the glycosylation will be primarily on the N-terminus of the protein, C-terminus of the protein, or both ends? Why?

5. I-cell disease is caused by defective lysosomal function. Cells isolated from individuals with I-cell disease produce normal levels of lysosomal acid hydrolases, but the enzymes are secreted from the cell. Speculate on some possible causes for this disease.

C. Extensions and applications

1. COP-I -coated vesicles mediate transfer of materials between the Golgi and ER. The formation of COP-I-coated vesicles depends on the regulated activity of COPI coatomer, Arf1, and ArfGAP1. Arf1 is a member of the Arf/Sar1p family of proteins discussed in section 27.7. Like all members of this family, Arf1 is active when bound to GTP and inactive when bound to GDP. Arf1-GTP binds to the membrane and mediates assembly of the COP-I coatomer proteins, triggering budding of the nascent vesicle (see figure 27.14 for an illustration). The COPI coatomer recruits ArfGAP1, a GTPase-activating protein that stimulates Arf1-GTP to hydrolyze its GTP, triggering disassembly of the coatomer as discussed in section 27.7.

One question that has intrigued researchers is how the timing of Arf-1 inactivation is controlled; if Arf1 hydrolyzes its GTP too early, the coatomer will disassemble before vesicle formation is complete. Previous studies have suggested that ArfGAP1 activity *in vitro* is greater on membranes composed of loosely-packed lipids compared to membranes composed of tightly-packed lipids. A recent paper by Bigay et al. (2003 *Nature* 426:563-566) further examined the process of Arf1 activation/deactivation by studying the kinetics of COP-I-coat formation on liposomes *in vitro* in the presence of purified COP-I coatomer, Arf1, and ArfGAP1. They used a series of different liposomes with the same lipid composition but different radii. Surprisingly, they discovered that liposome radius had a dramatic effect on the kinetics of Arf1 inactivation (as measured by GTP hydrolysis). Specifically, the rate of Arf1 inactivation increased as the size of the liposomes decreased, and peaked at a liposome radius of 30-40nm, which is typical of the size of COP-I-coated vesicles *in vitro*.

Using this information, postulate a possible mechanism for how the timing of Arf1 inactivation is controlled in COP-I-coated vesicles.

2. Cells infected by vesicular stomatitis virus (VSV) express the viral VSVG protein. VSVG protein is a group I single pass transmembrane protein that is inserted into the ER membrane as discussed in Chapter 8, and then transported through the Golgi and ultimately to the cell surface via constitutive secretion/exocytosis.

253

You are using VSVG protein as a marker to study protein trafficking through the secretory pathway. Your perform a series of experiments (as described below), then use immunocytochemistry to visualize where the VSVG protein is located. Predict the outcome for each experiment.

    a. Cells infected with VSV and treated with brefeldin A.

    b. Cells infected with a modified VSV that encodes a VSVG protein with a KKLL sequence at its C terminus.

    c. Cells infected with a modified VSV that encodes a VSVG protein with a KKLL sequence just after its N-terminal signal sequence.

D. Standardized test format

1. Which of the following does not take place in the ER?

    A. protein folding
    B. N-linked glycosylation
    C. oligomerization of multimeric proteins
    D. O-linked glyscosylation
    E. all the above take place in the ER

2. High mannose oligosaccharides are generated in the

    A. ER.
    B. *cis*-Golgi.
    C. medial-Golgi.
    D. *trans*-Golgi.
    E. nucleus.

3. COP-II-coated vesicles mediate which of the following transport processes?

    A. *trans*-Golgi to endosomal transport
    B. receptor mediated endocytosis
    C. ER to *cis*-Golgi transport
    D. *trans*-Golgi to *cis*-Golgi transport
    E. translation

4. Which of the following statements is true?

    A. Secretory vesicles form at the medial-Golgi.
    B. Secretory vesicles form from clathrin-coated vesicles.
    C. Adaptins form the outer layer of a clathrin-coated vesicle.
    D. Transmembrane proteins in clathrin-coated vesicles bind directly to clathrin.
    E. Secretory vesicles form from the nucleolar region within the nucleus.

5. What protein mediates the final separation of a budding vesicle from its site of origin?

    A. dynamin
    B. Arf
    C. Rab
    D. SNAP
    E. RNP

6. Which of the following is an accurate description of the exocyst's role in vesicular transport?

    A. It is tethering complex found on the vesicle membrane.
    B. It is a tethering complex found on the target membrane.
    C. It is a membrane fusion protein found on the vesicle membrane.
    D. It is a membrane fusion protein found on the target membrane.
    E. It is a membrane-bound organelle which transports lipids to the ER.

7. What is the function of ATP hydrolysis by NSF?

    A. to provide the energy for the initial binding of a v-SNARE to a t-SNARE
    B. to provide the energy necessary for membrane fusion
    C. to inactivate Rab protein prior to fusion
    D. to release the v-SNARE and t-SNARE complex after membrane fusion
    E. to remove the terminal phosphate from the NSF complex

8. Lysosomal proteins to which a KDEL sequence has been added are targeted to the ER instead of the Golgi. Why?

    A. The KDEL sequence blocks formation of clathrin-coated vesicles needed for transport to the lysosome.
    B. Mannose-6-phosphate receptors cannot bind to KDEL-containing proteins.
    C. KDEL receptors are found earlier in the Golgi stack than are the enzymes that mediate mannose-6-phosphate addition.
    D. KDEL receptors in the lysosome mediate formation of vesicles that return the protein to the via retrograde transport.
    E. KDEL receptors block all other transport functions in the cell.

9. What is the major function of the early endosome?

    A. It is the site of degradation of macromolecules taken into the cell by endocytosis.
    B. It is the site for sorting of proteins and other macromolecules taken into the cell by endocytosis.
    C. It is a site of protein sorting during constitutive secretion.
    D. It is the site for removal of mannose-6-phosphate oligosaccharides from lysosomal proteins.
    E. It couples with the ER for protein export.

10. Internalization signals on many proteins that participate in receptor-mediated endocytosis are small, cytoplasmic sequences that usually contain a _____ residue.

     A. threonine
     B. asparagine
     C. serine
     D. tyrosine
     E. nucleotide

Answers:

A. Recall and short answer

1. Dolichol is a lipid that serves as the site of assembly for the initial oligosaccharide group that is transferred to an asparagine residue during N-linked glycosylation.

2. Endo H is a glycosidase that cleaves the high mannose oligosaccharides generated in the ER, but not the inner core-containing oligosaccharides that form in the Golgi. Endo H susceptibility can therefore be used to track whether or not a glycoprotein has passed from the ER to the Golgi.

3. *cis*-Golgi

4. Coat proteins help create vesicles by deforming the donor membrane, causing it to "bud." In addition, the specific coat protein composition of a vesicle helps package the appropriate cargo into the vesicle.

5. retrograde; anterograde

6. Both ARF and dynamin are guanine nucleotide binding proteins that associate with the coatomer. ARF-GTP initiates the budding of coated vesicles, whereas dynamin-GTP is required for the final pinching off (scission) of a newly forming vesicle.

7. tethering complex; Rab; v-SNARE; t-SNARE

8. Lysosomal proteins have a mannose-6-phosphate residue added to them in the ER or early Golgi. This serves as a transport signal that directs mannose-6-phosphate-containing proteins into clathrin-coated vesicles forming at the *trans*-Golgi.

9. Early endosomes are located near the cell membrane, and serve as the initial sorting site for proteins internalized by endocytosis. Late endosomes are located deeper in the cell, and serve as intermediates between early endosomes and lysosomes.

10. The LDL-LDL receptor complex is brought to an endosome, where the two dissociate due to the low pH inside the endosome. The LDL particle is ultimately transported to lysosomes for degradation and release of the cholesterol. The LDL receptor is repackaged into vesicles and recycled back to the cell surface.

B. Analytical and critical thinking

1. One difference is in how the oligosaccharides are attached to the glycoprotein. In N-linked glycosylation, oligosaccharides are added to the amino (NH_2) group of asparagine. O-linked glycosylation occurs on the OH group of serine, threonine or hydroxylysine. Another difference is where in the cell the two glycosylation events occur. N-linked glycosylation begins on the ER and is finished in the Golgi, whereas O-linked glycosylation takes place entirely in the Golgi.

2. The coat proteins that assemble on the cytoplasmic side of a membrane during the initial steps of vesicle formation are required for the initial deformation of the membrane to produce a bud. However, these coat proteins must be removed from a vesicle prior to it fusing with its target membrane to allow the two membranes to come into close enough contact for fusion to occur.

3. The vesicular model of anterograde transport states that anterograde trafficking of proteins through the Golgi cisternae occurs primarily via vesicular transport; proteins get packaged into vesicles that pinch off of one cisterna and fuse with the next cisterna. In contrast, the cisternal maturation model proposes that entire cisternae mature from *cis-* to *trans-* via changes in their protein composition. In this model, the major role of vesicular transport between Golgi stacks is for the retrograde transport of Golgi-resident enzymes backwards after a cisterna matures.

4. The enzymes required for both N- and O-linked glycosylation are in the lumen of the ER and Golgi. Since group II transmembrane proteins are inserted into the ER with their N-terminus facing the cytoplasm, that half of the protein will never be exposed to the ER or Golgi environment. Therefore, only the C-terminal half of Lys1 will be glycosylated.

5. The fact that the acid hydrolases are produced at normal levels indicates that the defect is not in the lysosomal enzymes themselves or in their ability to enter the ER. The observation that acid hydrolases are secreted from the cells of patients with I-cell disease suggests that proper targeting of the enzymes to lysosomes is defective. The two most likely causes for this defect are that either the mannose-6-phosphate tag is not added in the Golgi, or that the mannose-6-phosphate receptor in the *trans*-Golgi is defective. In either case, the acid hydrolases would progress through the ER and get secreted via exocytosis rather than get targeted to the lysosomes. (In fact, most people with I-cell disease are missing GlcNAc phosphotransferase, and so fail to add the mannose-6-phosphate targeting signal to lysosomal proteins.)

C. Extensions and applications

1. Since the composition of the liposomes was the same, we can assume that Arf1 inactivation is not triggered by a change in the composition of the membrane. However, as the size of the liposomes decreases, the degree of curvature of their membranes will increase. One model to explain these results is that ArfGAP-1 has the ability to sense the curvature of a membrane, and when a certain degree of curvature is reached it stimulates GTP hydrolysis by Arf1. This would also explain the lipid-packing data; as a membrane curves, its lipids become less tightly packed and more spread out. If ArfGAP-1 has the ability to sense curvature/lipid density, then that would provide a very nice mechanism to ensure that coatomer disassembly does not occur prematurely.

2.
    a. Since brefeldin blocks anterograde transport by inhibiting the formation of COP-II vesicles, in this experiment the VSVG protein will accumulate in the ER/early Golgi.

    b. C-terminal KKXX is an ER retention signal for transmembrane proteins. Thus, this form of VSVG protein will be concentrated in the ER rather than get transported to the cell surface.

    c. In order for a KKLL sequence to function as an ER retention signal, the dilysine motif must interact with the COP-I coatomer on the cytoplasmic side of the membrane. The N-terminal end of a group I transmembrane proteins like the VSVG protein is lumenal, not cytoplasmic. Thus, an N-terminal KKLL sequence on VSV G protein will not function as an ER retention signal, and this form of the protein will be inserted into the plasma membrane normally.

D. Standardized test format

1. D
2. A
3. C
4. B
5. A
6. B
7. D
8. C
9. B
10. D

Chapter 28

Signal transduction

General Concepts	Section	Related Questions
Carriers and channels	28.1-28.4	A1, B1, B2, D1
G protein-linked receptors	28.1,28.5, 28.6	A2, B3, C2, D3, D9
Second messengers	28.5, 28.19	A10, C2, D2, D7, D9
Signaling through protein kinases	28.7, 28.8	A3
Receptor tyrosine kinases	28.8-28.16	A3, A4, B4, B6
Ras/MAP kinase signaling	28.11-28.17	A4, A6-A8, B4, B6, C1, D5, D8, D10
Protein-protein interaction	28.11-28.13	A5, A6, D4, D8
Amplification, specificity, and cross-talk	28.17,28.18,28.19	A9, B5
JAK-STAT signaling	28.20	B5, B6, D6
TGFβ signaling	28.21	B5, B6, D6

Questions:

A. Recall and short answer

1. Is the acetylcholine receptor an example of a ligand-gated ion channel or a voltage-gated ion channel?

2. What is the difference between the inactive and active forms of the α subunit of a trimeric G protein?

3. Protein kinases add a _____ group to a specific amino acid on their target proteins. Most cytosolic protein kinases are _____ (two words) kinases, whereas most receptor protein kinases are _____ kinases.

4. What is the first step in activation of a RTK after ligand binding?

5. SH2 domains allow proteins to bind to _____, whereas SH3 domains bind to proteins with the target amino acid sequence _____.

6. What structural properties of Grb2 allow it to function as an "adaptor" or "docking" protein?

7. During activation of the Ras/MAPK pathway, Ras activates the serine/threonine kinase _____, which in turn activates MEK.

8. Why do mutations in Ras that prevent it from being farnesylated prevent activation of the Ras/MAPK pathway?

9. What is meant by the term "crosstalk" during signal transduction?

10. The enzyme that generates cAMP is called _____ _____ (two words).

B. Analytical and critical thinking

1. What is the difference between a transporter and a channel?

2. In chapter 8 you learned that the membrane-spanning domains of most integral membrane proteins are composed of hydrophobic amino acids that form a hydrophobic α-helix. Does this rule apply to ion channels? Why or why not?

3. Briefly compare and contrast the activation of the α subunit of a trimeric G protein by ligand binding to a serpentine receptor with the activation of monomeric G proteins such as Ras upon ligand binding to a RTK.

4. In section 10.9 you were introduced to the concept of dominant negative mutations, which are mutations that create a form of a protein that has the ability to block the function of the wildtype form of the protein when both the mutant and wildtype proteins are present in the same cell. Mutations that delete all or most of the cytoplasmic domain of a RTK often result in a dominant negative form of the receptor, as shown below. Why do you think this mutant form will block the signaling activity of the wildtype form of the receptor?

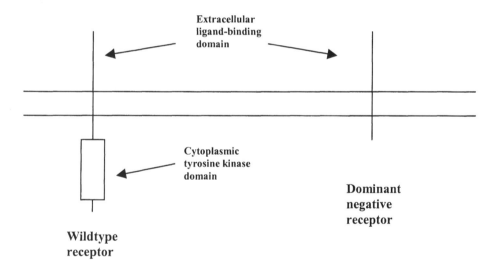

5. Briefly discuss the concept of amplification as it applies to signal transduction pathways. How do you think the Ras/MAPK, JAK/STAT, and TGFβ pathways compare in terms of their potential for amplification?

6. Briefly compare and contrast the activation of RTKs with the activation of TGFβ receptors.

C. Extensions and applications

1. The PDGF receptor (PDGFR) is a receptor tyrosine kinase whose ligand is a growth factor called PDGF. Autophosphorylation of PDGFR in response to ligand binding occurs on several different tyrosines in its cytoplasmic domain. You are interested in five of these in particular: Tyr^{740}, Tyr^{751}, Tyr^{771}, Tyr^{1009} and Tyr^{1021}. Three different SH2 domain-containing effectors are known to be recruited to these tyrosines after PDGFR autophosphorylation: PLCγ , the p85

subunit of PI3 kinase, and Ras-GAP. Figure 28.19 shows the structures of all three of these effectors.

       a. You are trying to figure out which effector binds to which phosphotyrosine. Your strategy is to create a series of PDGRF mutants in which a single tyrosine has been replaced by phenylalanine. You then express the mutant PDGFR in cultured cells, add PDGF to the culture medium, and compare the signal transduction response to that of PDGF-treated cells expressing the normal form of PDGFR and to that of cells that do not express PDGFR. Your data are as follows:

Cell line	PI3 levels	IP$_3$ levels	MAP kinase
No PDGFR	Very low	Very low	High
Normal PDGFR	High	High	Low
PDGFR with Tyr→Phe mutation at position 740	Very low	High	Low
PDGFR with Tyr→Phe mutation at position 751	Very low	High	Low
PDGFR with Tyr→ Phe mutation at position 771	High	High	High
PDGFR with Tyr→ Phe mutation at position 1009	High	Very low	Low
PDGFR with Tyr→ Phe mutation at position 1021	High	Very low	Low

       Using these data, create a model for how the three effectors are activated in response to PDGFR signaling.

       b. Surprisingly, upon reanalyzing the Tyr^{740} mutants you notice that not only are the levels of PI3 elevated to the same degree as in the positive controls, PI3 levels remain elevated for much <u>longer</u> in the Tyr^{740} mutant cells. Speculate as to why this might be. Hint: the amino acid sequence around the Tyr^{740} is Tyr-Met-Asp-Met (remember chapter 27!). (This question was inspired by Wu et al., 2003 *Journal Biological Chemistry* 278:40425-40428).

2. G$_s$-type trimeric G proteins activate adenylate cyclase, whereas G$_i$-type trimeric G-proteins <u>inhibit</u> adenylate cyclase (see figure 28.11). Both types of G proteins are activated by serpentine receptors, although different receptors activate G$_s$ or G$_i$ proteins. For example, binding of epinephrine to α$_2$-adrenergic receptors activates G$_i$, whereas binding of epinephrine to β$_2$-adrenergic receptors activates G$_s$. Cholera toxin (CT) is an enzyme that ribosylates the α subunit of G$_s$, thereby preventing it from hydrolyzing GTP. Pertussis toxin (PT) is an enzyme that ribosylates the α subunit of G$_i$, but in this case the ribosylation prevents G$_i$ from interacting with its appropriate serpentine receptors. Using this information, predict what effect each of the following experiments will have on intracellular cAMP levels compared to the same cells treated with epinephrine alone. Which treatment will result in the highest cAMP production?

<u>Experiment #1</u>: Treating cells that express both the α2- and β2-adrenergic receptors with epinephrine and with PT.

Experiment #2: Treating cells that express both the α2- and β2-adrenergic receptors with epinephrine and with CT.

Experiment #3: Treating cells that express both the α2- and β2-adrenergic receptors with epinephrine and with CT and PT.

D. Standardized test format

1. Which of the following statements about ion transport across a membrane is true?

   A. Carrier proteins always require ATP hydrolysis to facilitate active transport of ions.
   B. Ion channels facilitate the active transport of ions across a membrane.
   C. Ion channels are open all the time.
   D. Ion channels form a continuous water-soluble pore in the membrane.
   E. Ion channels form a continuous fat-soluble pore in the membrane.

2. Which of the following is not an example of a second messenger?

   A. cAMP
   B. IP_3
   C. DAG
   D. Ras
   E. All of the above are second messengers.

3. What is the major role of the β and γ subunits of a trimeric G protein?

   A. to prevent the α subunit from interacting with an effector by keeping it tethered to the receptor
   B. to cause the α subunit to release GDP and bind GTP
   C. to promote internalization of the receptor after ligand binding
   D. to bind to phosphotyrosines on the activated receptor
   E. to serve as an adaptor complex for tethering to the nuclear membrane

4. Which of the following statements is true?

   A. An SH2 domain-containing protein can bind to any phosphotyrosine.
   B. PTB domains are found primarily in docking proteins.
   C. WW domains bind to PDZ domains.
   D. SH3 domains bind to PPXY sequences.
   E. PTB domains function the same as WW and PDZ domains.

5. What effect would you predict on Ras/MAP kinase signaling in cells with a loss of function mutation in Ras-GAP?

   A. increased levels of IP_3
   B. inhibition of Raf
   C. prolonged activation of MAP kinase
   D. inactivation of GRB2
   E. "encapsulated" suppression of MAP kinase

6. What do STATs, SMADs, and CREB have in common?

    A. They are all activated by JAK kinases.
    B. They function as transcription factors when phosphorylated.
    C. They are regulated by second messengers.
    D. The signaling pathways that ultimately activate them are triggered by ligand binding to a receptor tyrosine kinase.
    E. The suppress the S phase of mitosis but not meiosis.

7. What is the major outcome of cAMP production by adenylate cyclase?

    A. activation of PLC
    B. activation of PKA
    C. activation of PI3 kinase
    D. inhibition of DAG
    E. degradation of PKA

8. What would be the most likely effect on the Ras/MAP kinase pathway of a mutation in the SH3 domain of Grb2?

    A. decreased autophosphorylation of the receptor tyrosine kinase
    B. the mutant Grb2 would not be able to bind to the activated receptor tyrosine kinase
    C. hyperactivation of Ras
    D. no recruitment of SOS to the membrane
    E. deposition of ten-fold amounts of SOS to the membrane

9. What would be the most likely effect on the transcription of CRE-containing genes in a cell expressing a constitutively-active form of the trimeric G_i protein?

    A. transient decrease in transcription, followed by a return to basal levels
    B. increased transcription over basal levels
    C. no effect
    D. prolonged decrease in transcription
    E. activation of all glycolytic pathways

10. Scaffold proteins such as the yeast STE5 protein function to…

    A. link Ras to Raf
    B. link STATs to JAKs
    C. co-localize MAPKKKs, MAPKKs, and MAPKs
    D. stimulate receptor internalization
    E. dispersal of MAPKKKs, MAPKKs, and MAPKs

Answers:

A. Recall and short answer

1. ligand-gated ion channel

2. The inactive form is bound to GDP. The active form is bound to GTP.

3. phosphate; serine/threonine; tyrosine

4. Ligand binding induces <u>dimerization</u> of the RTKs, which allows for autophosphorylation of their cytoplasmic domains.

5. phosphotyrosines; PXXP

6. One end of Grb2 has a SH2 domain, which allows it to bind to a phosphotyrosine on an activated RTK. The other end of Grb2 has an SH3 domain, which allows it to bind to SOS, the guanine nucleotide exchange factor that activates Ras.

7. Raf

8. Farnesylation allows Ras to be localized to the membrane. Ligand binding to a RTK recruits SOS to the membrane. If the Ras is not localized to the membrane because it is not farnesylated, then it cannot interact with SOS, and thus will remain GDP-bound and hence inactive.

9. Crosstalk refers to the phenomenon where a component in one pathway activates a signaling component in another pathway.

10. adenylate cyclase

B. Analytical and critical thinking

1. Channels form a continuous aqueous pore through the membrane that opens and closes in response to appropriate signals, allowing the appropriate ions to flow through the pore. Transporters, on the other hand, physically bind to and carry small molecules from one side of the membrane to the other by binding the ligand on the extracellular side of the membrane and then releasing it on the cytoplasmic side.

2. The membrane spanning domains of ion channels do form α-helices. However, these α-helices are often <u>amphipathic</u> rather than strictly hydrophobic. The hydrophobic side of each helix faces the membrane, whereas the hydrophilic side faces the center of the channel, creating an aqueous pore through which ions can travel across the membrane.

3. Both types of G proteins are inactive when bound to GDP, and active when bound to GTP. In both cases, ligand binding to the receptor ultimately causes in the G protein to switch from its inactive to active state. In the case of trimeric G proteins, this activation is <u>direct</u>; ligand binding to the receptor causes a conformational change in the receptor that in turn causes the α subunit to release GDP and bind GTP. In contrast, activation of monomeric G proteins by RTKs is <u>indirect</u>, meaning that Ras and other monomeric G proteins do not bind directly to the activated RTK.

Instead, ligand binding recruits GEFs (such as SOS) to the cell membrane, which in turn activate the monomeric G protein by triggering the release of GDP and acquisition of GTP.

4. The dominant negative form of the receptor has the ability to bind ligand (since the extracellular domain is unaffected), but lacks kinase activity. Thus, when this form of the receptor binds ligand it will still be able to dimerize, but will not be able to carry out the next step in activation, autophosphorylation. If the mutant form of the receptor is expressed at high enough levels in the cell, then most of the molecules of wildtype receptor will dimerize with a dominant negative receptor upon ligand binding, and these heterodimers will be unable to continue the signal transduction response.

5. Amplification of signal transduction pathways occurs because many of the steps in the signaling pathway are catalytic. In general, the more steps in a pathway, the more potential there is for amplification. For example, a single activated Ras can activate several molecules of Raf. Each Raf can then phosphorylate and activate many molecules of MAP kinase kinase, which in turn can phosphorylate and activate many different molecules of MAP kinase. In contrast to the Ras/MAP kinase pathway, the JAK/STAT pathway has very few steps; activated JAK kinase receptors directly phosphorylate the STATs, which then translocate to the nucleus and regulate transcription. Because there are fewer catalytic steps in the JAK/STAT pathway, there is less potential for amplification. The same holds true for TGFβ signaling, where the pathway is also much more simple and direct than the Ras/MAPK pathway.

6. Both RTKs and TGFβ receptors undergo multimerization upon ligand binding. However, whereas dimerization of RTKs triggers autophosphorylation of both receptors, only the type I TGFβ receptors gets phosphorylated in response to ligand-induced multimerization. In addition, RTKs are (as their name implies) tyrosine kinases, whereas the TGFβ receptors are Ser/Thr kinases.

C. Extensions and applications

1. a. PI3 and IP3 are second messengers generated by PI3 kinase and PLCγ, respectively. The lack of PI3 in Tyr^{740} <u>and</u> Tyr^{751} mutants indicates that PI3 kinase was not activated in those cells, suggesting that PI3 kinase normally binds to both of those phosphotyrosines. Similarly, the lack of IP_3 in Tyr^{1009} and Tyr^{1021} mutants suggests that the two SH2 domains in PLCγ bind to those two phosphotyrosines. GAP is a Ras-GTPase activating protein that helps <u>inactivate</u> Ras by stimulating GTP hydrolysis, thus antagonizing Ras activation by other receptors. In the Tyr^{771} mutants MAP kinase phosphorylation is increased, suggesting that Ras signaling has increased as a result of inactivation of GAP.

b. The fact that the IP3 levels stay high for a longer period of time in the Tyr^{740} mutants suggests that the normal downregulation of the receptor has been blocked by the mutation. Many activated receptors get downregulated by internalization via endocytosis. In chapter 27 you learned that YXXO (where O is any hydrophobic amino acid) acts as an internalization signal that triggers receptor-mediated endocytosis when the tyrosine is phosphorylated. The Tyr-Met-Asp-Met sequence around Tyr^{740} meets this criterion, and in fact Wu et al. demonstrated that Tyr^{740} phosphorylation is both necessary and sufficient to trigger receptor endocytosis.

2. In experiment #1, the PT will block activation of G_i but G_s will be unaffected. So, in the presence of epinephrine there will be higher than normal levels of cAMP. In experiment #2, the CT will cause G_s to be hyperactive (since it can no longer hydrolyze GTP to return to its inactive

state). G_i will still be active, but cAMP levels should still be higher than in cells treated with epinephrine only. Experiment #3 will result in the highest levels of cAMP production because G_i will be inhibited <u>and</u> G_s will be hyperactive.

D. Standardized test format

1. D
2. D
3. A
4. B
5. C
6. B
7. B
8. D
9. D
10. C

Chapter 29

Cell cycle and growth regulation

General Concepts	Section	Related Questions
Cell cycle checkpoints	29.2, 29.3, 29.13	A2, A6, A8, B1, B4, B5, D2, D3
Cyclins and cdks	29.5-29.7,29.9-29.11,29.14	A3, A4, B2, C1, D1, D5
Phosphorylation and cell cycle control	29.7, 29.12, 29.15	A4, A5, B2
Cdk substrates and inhibitors	29.16, 29.17	A6, B4, D6
Mitosis	29.18-29.29.24	A3, A7, A8, B1, B2, D7, D8
Apoptosis	29.25-29.30	A9, A10, B6, D9, D10
Techniques		
Cell fusions	29.4	B1
Yeast mutagenesis screens	29.8	B3

Questions:

A. Recall and short answer

1. For most cells, the longest phase of the cell cycle is _____. During this phase, the total DNA content of the cell is _____. Chromosome replication occurs during ____ phase. After this phase cells enter ___ phase, during which time their total DNA content is _____. Noncycling cells that have exited the cell cycle are said to be in _____.

2. What is the START point?

3. M phase kinase is a dimer composed of a _____ and _____. What role does M phase kinase play in the cell cycle?

4. What are the two general ways that M phase kinase activity is regulated?

5. Why are *wee1* mutants in yeast smaller than normal?

6. What are CKIs, and how do they function?

7. What types of cellular reorganizations are triggered at mitosis by the M phase kinase?

8. What is APC, and which cell cycle checkpoint is it involved in?

9. List three specific consequences of caspase activation inside apoptotic cells.

10. During apoptosis, _____ released by the mitochondria into the cytoplasm binds to APAF-1. This complex then activates _____, which in turn activates the effector caspase _____.

B. Analytical and critical thinking

1. Fusing a cell in G1 with a cell in M phase creates a heterokaryon in which the G1 nucleus immediately enters mitosis. This mitosis ultimately fails because the G1 cell's nucleus had not yet replicated its DNA. However, under normal conditions DNA replication is a prerequisite for mitosis, and mitosis will not begin until <u>after</u> the DNA replication checkpoint has been satisfied. Why does the G1 nucleus in the heterokaryon enter mitosis even though its DNA is unreplicated?

2. The B cyclins are one of the major mitotic cyclins in animal cells. Normally, B cyclin levels increase gradually during interphase and peak at the end of G2/beginning of M. If you overexpress cyclin B in cells during early interphase (for example G1), those cells do not immediately enter mitosis even though they are expressing the mitotic cdks (e.g., cdc2). Why?

3. Why do you think that the genetic screens in yeast (both *S. pombe* and *S. cerevisiae*) looked specifically for <u>temperature sensitive</u> cell cycle mutants, i.e., mutations that caused a phenotype at higher temperature, but had no effect at lower temperature? Why not just screen for mutations that showed a phenotype at all temperatures?

4. If you overexpress p27 in cells, what effect do you predict there would be on the phosphorylation of RB, and why?

5. Why are patients with the disease ataxia telangiectasia advised to avoid unprotected exposure to ultraviolet light or frequent X-rays?

6. All of the procaspases involved in apoptosis are "activated" by a common mechanism. What is it?

C. Extensions and applications

1. The dogma in mammalian cell cycle research has been that cdk2 activity is required for somatic cells to progress from G1 into S phase, first by association with cyclin E (to get past the restriction point), and then by association with cyclin A (to drive cells from G1 into S). This model is supported by a tremendous number of *in vitro* studies showing, for example, that inhibition of cdk2 with pharmacological agents or by injection of anti-CDK2 antibodies causes cells to arrest in G1. Recently, Ortega and colleagues (*Nature Genetics* 35:26-31) knocked out the *cdk2* gene in mice, and found that (i) Cdk2$^{-/-}$ mice are born viable and healthy; (ii) fibroblasts isolated from *Cdk2*$^{-/-}$ mice proliferate normally *in vitro*; and (iii) cyclin E immunoprecipitates from *Cdk2*$^{-/-}$ fibroblasts fail to phosphorylate histone H1 *in vitro*, whereas extracts from cyclin A immunoprecipitates of *Cdk2*$^{-/-}$ mice do phosphorylate histone H1 *in vitro*.

   a. Do these results support the idea that the G1/S transition is dependent on cdk2?
   b. Create a model to explain these results, and speculate about how *Cdk2*$^{-/-}$ mice might be able to compensate for lack of cdk2.

2. A trophic factor is an extracellular signaling molecule that activates a signal transduction cascade (see chapter 28) that protects cells from undergoing apoptosis. Many trophic factors work by regulating the phosphorylation of a cytoplasmic protein called BAD. BAD is a member of the Bcl2 family of proteins that, when unphosphorylated, binds to Bcl-2 and Bcl-X$_L$ and prevents them from translocating to the mitochondria. However, phosphorylation of BAD on Ser^{155} prevents BAD from associating with Bcl-2 or Bcl-X$_L$. In 2000, Zhou and colleagues

(*Journal of Biological Chemistry*, 275:25046-25051) demonstrated that treating cultured HeLa cells with the PKA inhibitor H89 blocked the phosphorylation of BAD on Ser[155], whereas treating cells with the adenylate cyclase-activating drug forskolin resulted in Ser[155] phosphorylation.

a. Which population of cells, forskolin-treated or H89-treated, do you predict is most likely to exhibit high levels of apoptosis, and why?

b. Based on these data and what you learned in chapter 28, speculate as to the likely nature of the trophic factor and the signaling pathway that regulates BAD Ser[155] phosphorylation.

D. Standardized test format

1. Which of the following statements about MPF is FALSE?

   A. It includes a cyclin and a cdk.
   B. It is active during all phases of the cell cycle.
   C. It was first described as an activity that promoted frog oocytes to enter meiosis.
   D. It is a cytoplasmic factor.
   E. All the above are true.

2. What does the cell "look for" at the G2 checkpoint?

   A. attachment of chromosomes to the mitotic spindle
   B. completion of DNA replication
   C. presence of extracellular growth factors
   D. adequate nutrient levels
   E. attachment of the centromere to the telomere

3. Which cell cycle transition is regulated by the checkpoint referred to as the "restriction point"?

   A. G1 → S
   B. S → G2
   C. G2 → M
   D. M → G1
   E. M → G2

4. At which phase of the cell cycle is a cell's total DNA content the greatest?

   A. late G1
   B. early G1
   C. G2
   D. early S phase
   E. the middle of G1

5. What is a key difference in cell cycle control between yeast and animal cells?

   A. Yeast only have one type of cyclin whereas animals cells have several.
   B. Yeast only have one type of cdk whereas animal cells have several.
   C. Phosphorylation of M phase kinase is not required in yeast.
   D. Phosphorylation of M phase kinase is not required in animal cells.
   E. Yeast do not have cyclins.

6. p16 is an example of a

    A. cyclin
    B. cdk
    C. checkpoint sensor protein
    D. cdk inhibitor
    E. cdk activator

7. Sister chromatids are held together by proteins called _____, and released during anaphase by the activity of proteins called _____.

    A. cohesins, separins
    B. separins, cohesins
    C. cohesins, securins
    D. securins, separins
    E. telomerases, isomerases

8. Which of the following statements about formation of the mitotic spindle is TRUE?

    A. It begins prior to nuclear envelope breakdown.
    B. It can be blocked by overexpression of RCC.
    C. It requires prior duplication of the centrioles during interphase.
    D. It requires inactivation of Ran.
    E. It requires that the centrioles duplicate four times during interphase.

9. What is the correct sequence of events in FasL-induced apoptosis?

    A. Aggregation of receptor → activation of caspase 3 → cytochrome c release → activation of caspase 8.
    B. Aggregation of receptor → activation of caspase 8 → cytochrome c release → activation of caspase 3.
    C. Aggregation of receptor → cytochrome c release → activation of caspase 3 → activation of caspase 8.
    D. Activation of caspase 8 → aggregation of receptor → cytochrome c release → activation of caspase 3.
    E. None of the above are correct.

10. What is the role of Diablo protein released by the mitochondrion during apoptosis?

    A. to bind to Apaf-1
    B. to help release cytochrome c
    C. to inhibit IAP proteins in the cytoplasm
    D. to bind to Bcl-2 and block its translocation to the mitochondria
    E. to activate cdk directly

Answers:

A. Recall and short answer

1. G1; 2C; S; G2; 4C; G0

2. The START point (or restriction point in animal cells) is the checkpoint in late G1 where cells make the decision whether or not to commit to DNA replication.

3. cyclin; cdc2. M phase kinase activity is required for cells to enter and complete mitosis.

4. One way is by the <u>levels of the cyclin</u> component (cyclins A and B, respectively). The cyclins gradually accumulate during interphase until they reach a peak at G2/M, and then disappear rapidly in late M as a result of proteolysis. The second way that M phase kinase activity is regulated is by the state of <u>phosphorylation of cdc2</u>. In order to be active, cdc2 must be dephosphorylated on Thr-14 and Tyr-15, and phosphorylated on Thr-161.

5. *wee1* encodes a kinase that phosphorylates Tyr-15 on cdc2, thus inactivating it. In the absence of Wee1 function cdc2 is hyperactive, so cells divide sooner than normal (before they have had time to double in size).

6. CKIs are cyclin-cdk kinase inhibitors. They bind to specific cyclin-cdk complexes and keep those complexes inactive, thus preventing progression through the cell cycle.

7. M phase kinase activity triggers the reorganization of the cell's microtubule cytoskeleton to form the mitotic spindle, breakdown of the nuclear membrane, condensation of the chromatin in chromosomes, and reorganization of actin microfilaments to make the contractile ring.

8. APC stands for anaphase promoting complex, an E3 ubiquitin ligase whose activity is required for mitotic cells to progress into anaphase and separate homologous chromosomes to opposite ends of the cell. APC activity is regulated by the Mad proteins, which inhibit the APC activator CDC20 unless all kinetochores are bound to spindle microtubules.

9. Activation of the apoptosis program results in DNA cleavage (and hence chromosome fragmentation), protein degradation, and damage to cell membranes leading to cell "blebbing" and eventual cellular fragmentation.

10. cytochrome c; caspase-9; caspases-3.

B. Analytical and critical thinking

1. The DNA replication checkpoint occurs in G2, prior to mitosis. Once this checkpoint has been satisfied, MPF becomes active and the cell enters mitosis. Since the mitotic cell had <u>already</u> passed through the DNA replication checkpoint <u>prior to</u> heterokaryon formation, it contained active MPF. This MPF was sufficient to trigger mitosis in the G1 nucleus.

2. Formation of the cdc2/cyclin B dimer alone is not sufficient to create active M phase kinase. cdc2 needs to have the proper pattern of phosphorylation (addition of an activating phosphate, loss of inhibitory phosphates) in order to form an active kinase. The necessary kinases and phosphatases are not active in G1, so the cdc2/cyclin B is not an active kinase complex.

272

3. Because *cdc* mutants fail to progress through the cell cycle, they cannot be propagated normally if they are expressing the mutant phenotype. This makes it very difficult to maintain stocks of the different mutant yeast strains. By isolating temperature sensitive mutants, researchers can propagate and maintain the mutant strains at low temperature, and then shift them to a higher temperature when they want to study the mutant phenotype. Also, by raising the temperature at different points in the cell cycle, researchers can more accurately ascertain the precise stage affected by the mutation.

4. p27 is a cyclin kinase inhibitor (CKI) that binds to and inhibits both cyclin D/cdk4,6 and cyclin E/cdk2. These cyclin/cdk complexes are required for entry into S phase, and act in part by phosphorylating RB. So, overexpression of p27 will prevent RB phosphorylation, and cells will not be able to pass from G1 into S.

5. Ataxia telangiectasia is caused by mutations in the *ATM* gene, which encodes a kinase that is activated at the G2 checkpoint by DNA damage. If ATM activity is absent or attenuated, many of the DNA repair proteins fail to get activated. Both ultraviolet and X-ray irradiation can cause DNA damage, and cells in patients with ataxia telangiectasia are less able to correct this damage, putting them at increased risk for developing cancer and other disorders.

6. Caspases are initially synthesized as inactive procaspases with a inhibitory N-terminal prodomain. The prodomain must be removed proteolytically in order for the caspase to become active.

C. Extensions and applications

1.

    a. The results of Ortega et al. demonstrate that, despite all the *in vitro* experiments to the contrary, mammalian cells *in vivo* do not depend on cdk2 to progress through the cell cycle. If they had, then the $Cdk2^{-/-}$ mice would have been early embryonic lethals and died as a result of defective cell proliferation during embryogenesis.

    b. The simplest explanation for these findings is that other cdk/cyclin combination are able to compensate for the lack of cdk2. As figure 29.27 shows, early in G1 there are at least 3 different cyclin/cdk combinations present in cells: cdk2/cyclinD, cdk4/cyclinD, and cdk5/cyclinD. Therefore, perhaps cdk4 and cdk5 are able to compensate for lack of cdk2 at this early stage. The cdk2/cyclinE complex shown later in G1 is probably NOT important for the G1/S transition *in vivo*, since the cyclin E immuoprecipitation/histone H1 phosphorylation assay demonstrates that cyclin E does not form a functional kinase complex with any other cdk in the $Cdk2^{-/-}$ mice. However, the fact that cyclinA immunoprecipitates <u>do</u> have H1 kinase activity in $Cdk2^{-/-}$ cells suggests that there is another cdk present in late G1/early S that is able to make a functional complex with cyclin A.

2.

    a. Bcl-2 and Bcl-X$_L$ are anti-apoptotic members of the Bcl2 family that work by interfering with pro-apoptotic Bcl2 family members such as Bax and Bcl-X$_S$. This interaction prevents the release of cytochrome c from the mitochondria, and therefore prevents apoptosis. By binding to Bcl-2 and Bcl-X$_L$ and preventing them from going to the mitochondria and inhibiting Bax and Bcl-X$_S$, the unphosphorylated form of BAD will promote apoptosis. Thus, the forskolin-treated cells should be resistant to apoptosis and the H89-treated cells should exhibit high levels of apoptosis.

273

b. PKA is activated by cAMP, which is generated by the the effector enzyme adenylate cyclase. Adenylate cyclase is most commonly activated by G_s –type trimeric G proteins. So, the trophic factor that regulates BAD phosphorylation likely acts via a serpentine, G_s-coupled receptor.

D. Standardized test format

1. B
2. B
3. A
4. C
5. B
6. D
7. A
8. C
9. B
10. C

Chapter 30

Oncogenes and cancer

General Concepts	Section	Related Questions
Immortalization and transformation	30.1, 30.2, 30.25, 30.26	A1, B2, C2
Tumor viruses	30.4-30.7, 30.11	A3
Proto-oncogenes and oncogenes	30.3, 30.8, 30.9-30.18	A2, A4-A6, A9, B3
Tumor suppressor genes	30.3, 30.19 - 30.23	A7, A8, B1, B4, C1
Aging and senescence	30.24, 30.27	A10, B1
Genetic instability	30.28, 30.29	A5, B1, B5, C2

Questions:

A. Recall and short answer

1. Describe 3 differences in the *in vitro* growth properties of cancer cells compared to normal cells.

2. What is the difference between a proto-oncogene, an oncogene, and a tumor suppressor gene?

3. Infection by a tumor virus is one way in which a normal cell can become transformed. T antigen is an example of an oncogene carried by a _____ _____ _____. *v-src* is an example of an oncogene carried by a _____ _____.

4. True or false. All transforming mutations in a proto-oncogene occur in the coding region, and hence change the sequence of the protein encoded by the resulting oncogene. Explain.

5. The *Philadelphia* chromosome seen in patients with chronic myelogenous leukemia is the result of a _____ between chromosomes 22 and 9 that creates a fusion between the _____ and _____ genes.

6. Oncogenic mutations in receptor tyrosine kinases that function as growth factor receptors usually promote the ligand-independent _____ of the receptor, which results in constitutive activation.

7. Is *RB* considered a proto-oncogene or a tumor suppressor gene, and why? How do certain viral oncoproteins impact the RB pathway?

8. The *INK4A-ARF* locus encodes a transcript that generates two different proteins as a result of alternative splicing. One of these proteins, p16^{INK4A} inhibits the kinase that normally phosphorylates the tumor suppressor _____. The other protein, p19^{ARF}, binds to and antagonizes the p53 inhibitor _____.

9. Give an example of an oncogenic mutation in *c-Ras*.

10. What is replicative senescence, and how does it relate to telomeres?

275

B. Analytical and critical thinking

1. Why is p53 sometimes referred to as the "guardian of the genome"?

2. The ability of many oncogenes to transform cells *in vitro* depends on whether the cell being transfected is from a primary cell culture or an immortalized cell line. For example, oncogenic forms of *c-Ras* readily transform 3T3 cells in culture, but not primary cultures. Why?

3. Predict the oncogenic potential of the following *Src* mutants, and briefly explain your answer.
   a. A point mutation that converts the tyrosine at position 527 to phenylalanine
   b. A deletion of most of the SH2 domain
   c. A deletion of amino acids 1-15 *and* a point mutation that converts the tyrosine at position 527 to a phenylalanine.

4. Define the term *dominant negative mutation* and give an example of a dominant negative mutation in a tumor suppressor gene that has transforming activity.

5. Why do patients with hereditary nonpolyposis colorectal cancer (HNPCC) typically exhibit abnormal microsatellite repeat lengths? Do you think that these abnormal microsatellites are the direct cause of the cancer phenotype?

C. Extensions and applications

1. For several years it has been known that p53 can induce apoptosis in two ways: a transcription-dependent pathway (e.g., through downregulation of genes whose protein products normally inhibit apoptosis), as well as by a transcription-independent pathway. A recent study published in *Molecular Cell* (Mihara et al., 2003, 11:577-590) sheds light on the transcription-independent pathway by demonstrating that (a) p53 relocates to the mitochondrial membrane in response to DNA damage, (b) artificially targeting p53 to the mitochondrial membrane triggers apoptosis, (c) antibodies to p53 coimmunoprecipitate Bcl-2 from radiation-damaged cells, and (d) a form of p53 lacking amino acids 239-248 is unable to bind to Bcl2 in immunoprecipitation assays.
   a. Use this information as well as the information in sections 30.20-30.23 of this chapter and 29.27-29.29 of chapter 29 to create a model for how p53 might trigger apoptosis in a transcription-independent fashion.
   b. Many naturally occurring mutations in p53 found in tumors occur in the DNA binding domain of p53. Would you expect these mutations to impact both of the p53-triggered apoptosis pathways? Why or why not?

2. Many drug companies are racing to design small molecules that target the oncogenic forms of specific proteins without affecting the proto-oncogenic form of the protein. One apparent early success was the drug Gleevac, which specifically inhibits the oncogenic Bcr-Abl fusion protein by binding to its kinase domain. Abl is a tyrosine kinase that promotes proliferation of hematopoeitic cells, and the Bcr-Abl fusion protein creates a hyperactive form of Abl. In early studies of Gleevac, patients with chronic myelogenous leukemia showed an initial positive response to Gleevac, but then after several weeks experienced a relapse of the disease. This phenomenon is particularly common in patients that started Gleevac treatment relatively late in the disease. Given what you have learned about cancer progression, why do you think Gleevac-treated patients eventually relapsed?

D. Standardized test format

1. What is a mutator gene?

    A. a gene whose normal function is to promote cell proliferation
    B. a gene whose normal function is to repair damaged DNA
    C. the oncogenic form of a proto-oncogene
    D. a retroviral-derived gene that causes a mutation when it inserts into the host cell's chromosome
    E. a proto-oncogene with normal activity

2. Choose the correct order of tumorigenic changes in cultured cells.

    A. primary cells→crisis→ transformation→immortalization
    B. primary cells→transformation→ crisis→ immortalization
    C. primary cells→immortalization→ crisis→ transformation
    D. primary cells→ crisis→ immortalization→ transformation
    E. secondary cells → immortalization → transformation → primary cells

3. Identify the true statement regarding permissive and nonpermissive cells (in the context of viral infection).

    A. Viruses cannot infect nonpermissive cells, but they can infect permissive cells.
    B. Infection of a nonpermissive cell by a DNA tumor virus causes that cell to die as a result of lysis.
    C. DNA tumor viruses can only transform nonpermissive cells.
    D. DNA tumor viruses can transform both permissive and nonpermissive cells.
    E. Permissive cells are proto-oncogenic while nonpermissive cells are oncogenic.

4. Which of the following is not a proto-oncogene?

    A. *Rb*
    B. *Ras*
    C. *Src*
    D. *Myc*
    E. All the above are proto-oncogenes.

5. Which of the following statements is true?

    A. All oncogenic mutations affect the amino acid sequence of the protein.
    B. Constitutive activation of a tumor suppressor gene is a transforming event.
    C. Both copies of a tumor suppressor gene must acquire a mutation in order for
        transformation to occur.
    D. Amplification of tumor suppressor genes occurs in many tumors.
    E. All proto-oncogenes are mutant forms of oncogenes.

6. B-cell tumors often show translocations between the _____ gene on chromosome 8 and the IgH locus on chromosome 14.

    A. *c-myc*
    B. *c-ras*
    C. *c-src*
    D. *c-Abl*
    E. *tribble*

7. Transforming mutations in receptor tyrosine kinase-type growth factor receptors can occur as a result of what kind of change?

    A. deletion of the kinase domain
    B. deletion of the ligand binding domain
    C. point mutations in the cytoplasmic tail outside of the kinase domain
    D. a single point mutation in the ligand binding domain
    E. none of the above

8. Which of the following statements about RB is FALSE?

    A. It controls entry of cells into S phase.
    B. It functions as a transcriptional repressor when unphosphorylated.
    C. It binds to E2F when phosphorylated.
    D. It is a tumor suppressor gene.
    E. All the above are true.

9. Which of the following statements is TRUE?

    A. Telomerase-negative mice develop tumors faster than p53-negative mice.
    B. T-antigens of DNA tumor viruses bind either to p53 or RB, but not both.
    C. Telomerase activity must be restored for cells to survive crisis.
    D. p53 binds to intact telomeres.
    E. *p53* is an oncogene.

10. Which of the following is <u>not</u> likely to lead to increased genetic instability?

    A. continued cell division after telomere loss
    B. loss of function mutations in both alleles of p53
    C. mutations in mismatch repair systems
    D. deletion of the *Mdm2* gene
    E. none of the above

Answers:

A. Recall and short answer

1. Any 3 of the following are acceptable.

   i. Most normal cells require a solid surface on which to attach in order to grow and survive in culture, a phenomenon called <u>anchorage dependence</u>. Many tumor cells are capable of growing in an anchorage-independent fashion, meaning that they can survive and grow in suspension or when cultured in/on non-adhesive substrata.

   ii. Normal cells require the addition of exogenous growth factors in order to grow in culture, and serum is the most commonly used source of these growth factors for most cell cultures. Tumor cells, however, often grow and survive just fine in culture without serum or any other outside source of growth factors.

   iii. Normal cells stop dividing when they have covered the available surface area of the growth substrate and do not pile up on top of each other. Many tumor cells, however, do not exhibit this contact-dependent inhibition.

   iv. Normal cells tend to be flatter and more elongated than tumor cells, which are often rounded up. This is due in part to the fact that tumor cells tend to have a much less organized cytoskeleton than do normal cells.

   v. Transformed cells are immortal and will proliferate forever, whereas normal cells die after a finite number of cell divisions.

2. Proto-oncogenes are genes whose protein products normally promote cell growth. They are converted to their cancer-causing form, called an oncogene, by gain-of-function mutations. Tumor suppressor genes are genes whose protein products normally slow cell proliferation. Transforming mutations in tumor suppressor genes are loss-of-function mutations.

3. DNA tumor virus; tumor retrovirus

4. False. Some oncogenic mutations result in the excessive production of a proto-oncogenic protein and do not affect the amino acid sequence of that protein. For example, extra copies of *myc*, either as a result of infection by a *v-myc*-carrying retrovirus or amplification of the *c-myc* locus, can be tumorigenic even if the protein encoded by the extra copies of *myc* has a normal amino acid sequence.

5. translocation; *Bcr*; *c-Abl*

6. dimerization

7. *RB* encodes a protein that, when unphosphorylated, slows down cell cycle progression by inhibiting the entry of cells into S phase. It is considered a tumor suppressor gene because loss-of-function mutations in <u>both</u> alleles of *RB* are required for tumors to arise. Some tumor viruses encode proteins that bind to and inhibit RB. For example, the SV40 T antigen and the adenovirus E1A proteins both bind to unphosphorylated RB and prevent it from binding to (and inhibiting) the transcription factor E2F. As a result, cells progress out of G1/ (or G0) and into S phase inappropriately.

8. RB; Mdm2

9. Ras proteins are all in their active state when bound to GTP. A mutation that reduces or eliminates the ability of a Ras to hydrolyze GTP (either by inactivating the intrinsic GTPase activity of Ras or by preventing Ras from associating with GAPs) will cause the Ras protein to be constitutively active.

10. Replicative senescence describes the phenomenon whereby normal cells in culture die rapidly after a certain number of divisions. There is evidence that this is triggered, at least in part, by the loss of telomeres in cells that have undergone many divisions. Loss of telomeres results in the activation of p53, which triggers apoptosis.

B. Analytical and critical thinking

1. p53 acts as a guardian of the genome because loss of p53 increases genomic instability in two ways. First, p53 gets activated in response to DNA damage, and in turn activates some of the DNA repair systems in cells (e.g. GADD45). Second, p53 triggers apoptosis in response to severe DNA damage or loss of telomeres, which prevents the damaged chromosomes from being replicated aberrantly and contributing to progressive genetic instability in later generations.

2. Primary cell cultures are relatively "normal" and still possess all or most of the pathways that act to limit uncontrolled cell cycle progression. Therefore, although they might exhibit higher rates of proliferation when transfected with an oncogene, they generally cannot proliferate indefinitely and undergo cell death when their telomeres get too short or they accumulate excessive DNA damage, etc. In contrast, immortalized cell lines have passed through the crisis point already, and so have lost many of the normal tumor suppressor pathways (e.g., loss of p53, activation of telomerase, etc). As a result, when these cells begin to express an oncoprotein, there are few (if any) constraints on their ability to proliferate indefinitely, and so they ultimately form tumors.

3.

    a. Phosphorylation of tyrosine 527 inactivates c-Src. Converting that tyrosine to a phenylalanine removes this potential site of negative regulation, and so the protein is more active (and thus highly oncogenic).

    b. The SH2 domain of c-Src allows it to interact with phosphotyrosines on an activated receptor tyrosine kinase. Without the SH2 domain, Src cannot associate strongly with the receptor, and so will not get activated very efficiently in response to ligand binding. This form of c-Src will not be very oncogenic.

    c. This form of c-Src will be only weakly oncogenic. Conversion of tyrosine 527 to a phenylalanine normally results in oncogenic activation of the protein, as discussed in (a). However, the N-terminal domain of c-Src is the site of myristylation, which is required to localize c-Src to the cell membrane, where it is believed many of the targets of Src are located.

4. A dominant negative mutation creates a form of a protein that interferes with the function of the normal version of that same protein. As a result, even if the individual has one normal copy of the gene, they still express the mutant phenotype. For example, p53 functions as a tetramer in the cell. A mutation in the DNA binding domain of p53 renders it unable to activate transcription. Tetramers composed of a mixture of normal p53 and p53 with an abnormal DNA binding domain are also nonfunctional.

5. Tumor cells derived from patients with HNPCC have mutations in genes involved in the DNA mismatch-repair systems. As a result, these cells have a higher mutation rate than normal cells. One potential consequence of this higher mutation rate that is easily assayed is decreased microsatellite repeat lengths, although these microsatellite changes in and of themselves do not cause the tumorigenic phenotype.

C. Extensions and applications

1.

a. Bcl-2 is an anti-apoptosis protein found at the mitochondrial membrane that prevents cytochrome c release into the cytoplasm. If p53 binds to Bcl-2 and prevents it from carrying out its anti-apoptotic function, then targeting of p53 to the mitochondria in response to DNA damage, etc., will result in the release of cytochrome c into the cytoplasm, which will in turn activate the caspase cascade that ultimately brings about apoptosis.

b. The DNA binding domain of p53 extends from about amino acid 100-290 (see figure 30.34). This region also includes the region implicated in binding to Bcl-2. Therefore, mutations to many parts of the DNA binding domain will likely also abrogate the ability of p53 to trigger apoptosis via the mitochondria.

2. As cancer cells progress, they become more and more genetically unstable through loss of p53, mutations in DNA repairs systems, etc. The result is that cancer cells accumulate increasing numbers of mutations over time. In this specific example, you might predict that patients eventually acquire mutations in the *Bcr-Abl* fusion gene that prevent the Bcr-Abl protein from interacting with Gleevac. In fact, this was demonstrated directly in a 2002 study by Shah et al. (*Mol. Cell*, 2:117-125), in which it was shown that 29 out of 32 CML patients who had relapsed after Gleevac treatment had acquired mutations in the kinase domain of Bcr-Abl.

D. Standardized test format

1. B
2. D
3. C
4. A
5. C
6. A
7. B
8. C
9. C
10. D

Chapter 31

Gradients, cascades, and signaling pathways

General Concepts	Section	Related Questions
Maternal genes, morphogens, and gradients	31.2-31.13	A1-A7, B1, B2, C1, D1-D3
Segmentation genes	31.14-31.19	A10-A12, D5-D7, D10
Homeotic genes	31.20-31.22	A14, A15, B4, B5, C2, D9
Signaling pathways during development	31.9, 31.10, 31.13, 31.18	A8, A9, A13, B3, D4, D7, D8, D10

Questions:

A. Recall and short answer

1. Briefly compare and contrast *Drosophila* **maternal somatic genes** and **maternal germline genes**. What type of phenotype results from mutations in these classes of genes?

2. A _____ is a protein whose specific local concentration dictates cell fate or regional identity in the embryo. _____ fulfills this role in the anterior patterning system of *Drosophila*.

3. How do bicoid and nanos regulate *hunchback* expression?

4. The activation of _____ by gurken on dorsal follicle cells results in the activation of a Ras signaling pathway that ultimately prevents inappropriate activation of the _____- determining pathway.

5. Why is it that, even though the Toll protein is expressed uniformly around the embryo, it is only activated on the ventral side?

6. What phenotype would you expect from a mutation in *cactus* that prevents the cactus protein from binding to the dorsal protein?

7. Briefly describe the two different ways that the initial axial asymmetry can be established in the egg.

8. The product of the *Drosophila* gene *dpp* is a member of which family of signaling molecules?

9. Dpp and its vertebrate homolog, BMP4, both act to _____ the formation of neural tissue; however, whereas dpp is active on the _____ side of *Drosophila* embryos, BMP4 is active on the _____ of frog embryos.

10. What are the 3 classes of segmentation genes, and in what order do they act?

11. What phenotype would you expect from a loss of function mutation in *ftz*?

282

12. True or false: all of the segment polarity genes encode transcription factors.

13. What will happen to armadillo (β-catenin) levels if Zw3 (GSK3) is inactivated?

14. The homeotic genes that specify the anterior parts of the fly are located in the _____ complex, whereas those that specify the posterior thorax and abdomen are found in the _____ complex.

15. Define the terms **homeotic gene**, **homeobox**, and **homeodomain**.

B. Analytical and critical thinking

1. Cytoplasm is removed from the posterior side of an early wildtype *Drosophila* embryo and injected into the anterior end of another wildtype fertilized egg. Predict the phenotype of the recipient embryo, and explain your reasoning.

2. *bicoid* codes for a transcription factor that acts as a morphogen to establish anterior-poterior identity during early *Drosophila* development. However, the morphogens that act later in development, such as dpp, are secreted signaling molecules. Would you expect that transcription factors could serve as morphogens in later development, or conversely that secreted signaling molecules could act as morphogens in early development? Why or why not?

3. Pretend that you are a scientist who has just completed a screen for mutations affecting neural development in *Drosophila*. Two of the more interesting mutations that you discover are both in the *sog* gene, but they give completely opposite phenotypes. sog^A mutants are dorsalized (meaning that they have an expansion of mesenchymal structures at the expense of neural structures), whereas sog^B mutants are ventralized (meaning that they have extra neural structures at the expense of dorsal structures). Speculate as to what kinds of mutations might have occurred in each allele.

4. What is the distinction between a conventional gene and complex loci such as *ANT-C* and *BX-C*?

5. The rule of "posterior prevalence" states that more posteriorly-expressed homeotic genes tend to be "dominant" over homeotic genes that are expressed more anteriorly. Explain how this property relates to *Ubx* function in *Drosophila*.

C. Extensions and applications

1. Bicoid is a homeodomain-containing transcription factor that regulates the expression of several genes, including *hunchback*. In addition, however, bicoid is also a capable of inhibiting the <u>translation</u> of *caudal* mRNA. *caudal* mRNA is found uniformly along the anterior-posterior axis of *Drosophila* embryos, but the caudal protein is localized in a sharp posterior-anterior gradient. Bicoid binds to the 3' UTR of *caudal* mRNA, and Niessing et al. (*Genes and Development*, 16: 2576-2582) determined that a YIPRL motif in the amino half of the bicoid protein is capable of binding to and interfering with the translation initiator protein eIF-4E (Chapter 6). A diagram of the bicoid protein is shown below.

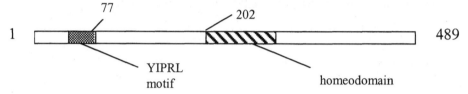

Below are distributions of caudal protein (left) and <u>zygotic (not maternal)</u> *hunchback* mRNA (right) in wildtype embryos. Anterior is to the left and posterior is to the right of each embryo.

What do you predict these gradients will look like in the following genotypes, and why?
   a) *Bcd^{null}* embryos (complete deletion of the *Bcd* locus)
   b) *Bcd^{77-489}* embryos (deletion in *Bcd* that removes amino acids 1-77)
   c) *Bcd^{1-202}* embryos (deletion in *Bcd* that removes amino acids 202-489)
   d) *Bcd^{3X}* embryos (transgenic flies that express 3 extra copies of *Bcd* under the control of the normal *Bcd* promoter)

2. Although vertebrate development is quite different than *Drosophila* development, there are many similarities. Just as adult *Drosphila* segments are derived from the embryonic parasegments, segmented structures in adult vertebrates, such as the vertebrae, are derived from segmented embryonic structures, in this case somites. The homeotic genes in the *ANT-C* and *BX-C* complexes control the proper anterior-posterior patterning of *Drosophila* segments. Likewise, clusters of vertebrate homeotic genes called Hox genes (analogous to, and derived from, the *ANT-C* and *BX-C* complexes in *Drosophila*) control proper anterior-posterior patterning of the vertebrae. One of the vertebrate Hox gene clusters is called the Hox A cluster. In mammals, this cluster has 11 different homeobox-containing genes arranged as follows:

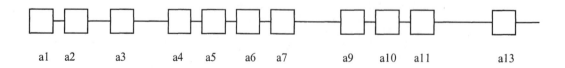

Pretend that you are a scientist studying the Hox A cluster in a newly discovered rodent. The expression patterns of some of the *Hoxa* genes in anterior somites of embryos from this rodent are given below, relative to what type of vertebra the somite gives rise to. For example, C1 = the somite that gives rise to the first cervical vertebra, T1 = the somite that gives rise to the first thoracic vertebra, etc. Your studies have also shown that, in contrast to other mammals, no Hox B, Hox C, or Hox D genes are expressed in these anterior somites.

C1 C2 C3 C4 C5 C6 C7 T1 T2 T3 T4 T5 T6 T7 T8...

Hoxa1 expression

Hoxa2 expression

Hoxa4 expression

Use the information provided above and what you know about how genes in the *ANT-C* and *BX-C* complexes pattern fly segments to answer the following questions.

a. Where do you think *Hoxa3* will be expressed, and why?
b. What phenotype would you expect if you knocked out the *Hoxa2* gene, and why?
c. What phenotype would you expect if you made a transgenic embryo that misexpressed *Hoxa4* under the control of the *Hoxa2* promoter, and why?

D. Standardized test format

1. *nanos* is an example of a(n) _____.

    A. maternal gene
    B. segmenation gene
    C. homeotic gene
    D. anterior gene
    E. distal gene

2. Which of the following is NOT a morphogen?

    A. bicoid
    B. nanos
    C. dpp
    D. hunchback
    E. All the above are morphogens.

3. What is the role of Gurken and Torpedo in dorsoventral signaling?

   A. To repress *Toll* expression in dorsal follicle cells.
   B. To promote *Toll* expression in ventral follicle cells.
   C. To repress spatzle activity in the dorsal side of the embryo.
   D. To activate dorsal in the ventral region of the embryo.
   E. Neither has a role in dorsoventral signaling.

4. Dorsal and cactus are homologues of which two vertebrate proteins?

   A. NF-κB and I-κB
   B. BMP and TGFβ
   C. Noggin and chordin
   D. Smad1 and Smad4
   E. p53 and E2F

5. Which of the following statements about gap genes is true?

   A. They are expressed in every other segment.
   B. They are expressed during oogenesis.
   C. They directly activate the expression of homeobox genes.
   D. Different gap genes regulate each other's expression.
   E. All are dependent on formation of the ovarian cortex.

6. What is the phenotype of a *Drosophila* embryo with a loss of function mutation in *eve*?

   A. loss of all head structures
   B. dorsalization of the embryo
   C. loss of odd-numbered parasegments
   D. loss of even-numbered parasegments
   E. loss of all abdominal segments

7. What maintains *engrailed* expression at segment boundaries?

   A. *Dpp* expression by a neighboring cell
   B. *Wg* expression by a neighboring cell
   C. *Hh* expression by a neighboring cell
   D. *Sog* expression by a neighboring cell
   E. All the above are correct.

8. Identify the TRUE statement regarding the role of Dsh during wingless/wnt signaling.

   A. It is the receptor for wingless/wnt.
   B. It phosphorylates Arm/β-catenin.
   C. It inhibits Zw3/GSK3.
   D. Loss of function mutations cause ectopic wingless/wnt signaling.
   E. None are true.

9. Which of the following statements about the *ANT-C* is FALSE?

    A. It contains genes involved in the formation of head segments.
    B. It is an example of a complex locus.
    C. It contains few protein-coding genes but many *cis*-acting regulatory sites.
    D. Mutations in *ANT-C* genes usually result in homeotic transformations.
    E. none of the above

10. What phenotype would you expect in an embryo with a null mutation in both alleles of *hedgehog*?

    A. loss of very other segment
    B. homeotic transformations
    C. loss of anterior structures, no effect on posterior structures
    D. loss of A/P compartment identity in segments
    E. loss of legs in the adult fly

Answers:

A. Recall and short answer

1. Both maternal somatic genes and maternal germline genes are expressed prior to fertilization, and mutations in both types of genes result in female sterility. The difference between them relates to <u>where</u> they are expressed: maternal germline genes are expressed in nurse cells or the oocyte. Maternal somatic genes also affect egg development, but are expressed in somatic tissues (the follicle cells).

2. morphogen, bicoid

3. Bicoid acts as a transcriptional activator of *hunchback,* whereas nanos acts as a translational repressor of *hunchback* mRNA. Since bicoid is at its highest concentration anteriorly and nanos is at its highest concentration posteriorly, the result is very high levels of hunchback protein in the anterior of the embryo, but essentially no hunchback protein in the posterior of the embryo.

4. torpedo, ventral

5. Toll encodes a transmembrane receptor whose ligand is spatzle. Since spatzle is only proteolytically activated in the perivitelline space on the ventral surface of the egg, only the Toll protein on the ventral egg membrane gets activated.

6. Because cactus normally binds to dorsal and prevents dorsal from translocating to the nucleus, a mutation that blocks the interaction between dorsal and cactus will result in excessive nuclear dorsal, and therefore a <u>ventralized</u> embryo.

7. In the anterior and posterior patterning systems, the initial asymmetries are established by the localization of specific mRNAs, which in turn create protein gradients. In the dorsal-ventral and terminal systems, the initial asymmetry is a result of the localized activation of a receptor protein, leading to the activation of a signal transduction pathway only in that part of the embryo.

8. the TGFβ family

9. inhibit, dorsal, ventral

10. <u>Gap genes</u> are activated by maternal patterning genes. The gap genes then activate the <u>pair rule</u> genes, which in turn activate the <u>segment polarity genes</u>.

11. *ftz* is a pair-rule gene that is expressed in the even numbered parasegments of *Drosophila* embryos. Mutations in *ftz* will result in those parasegments not forming, so the embryo will have half the normal number of segments.

12. False. Although some segment polarity genes do encode transcription factors (e.g., *engrailed*), others encode secreted signaling molecules (e.g. *wingless, hedgehog*), receptors (e.g., *patched*), and cytoplasmic signal transduction effectors (e.g. *disheveled*).

13. Zw3/GSK is a kinase that phorphorylates Arm/β-catenin, causing it to be degraded in the cell. In the absence of active Zw3, Arm/β-catenin will <u>not</u> be phosphorylated so its levels will increase, allowing it to enter the nucleus and regulate gene expression.

14. ANT-C, BX-C

15. Homeotic genes are genes that, when mutated, result in a transformation of segment identity, rather than a simple loss of a segment. A homeodomain is a ca. 60 amino acid-long DNA binding domain found in certain transcription factors. The homeobox is the region of a gene that encodes a homeodomain.

B. Analytical and critical thinking

1. The posterior cytoplasm contains the *nanos* RNA, so injecting posterior cytoplasm to the anterior side of another embryo will result in the ectopic expression of nanos protein anteriorly. Since nanos represses the expression of *hunchback*, which is needed for formation of anterior structures, the result will a posteriorized embryo (loss of anterior structures).

2. The early *Drosophila* is a syncytial blastoderm, meaning that all of the nuclei are located in a common cytoplasm (no cytokinesis after karyokinesis). This allows intracellular factors such as a transcription factors to diffuse throughout this common cytoplasm and form a morphogen gradient. It also means that there are effectively no cell-cell interactions at these stages, and so no real mechanism for cell-cell communication via secreted signaling molecules. Later embryos, in contrast, are completely cellularized and so transcription factors cannot diffuse from one end of the embryo to another; they are trapped in the cell in which they are produced. This means that the only way to establish a morphogen gradient in older embryos is by the localized production and secretion of diffusible extracellular signaling molecules.

3. Dpp inhibits neural (ventral) differentiation and promotes differentiantion of ventral structures. Since the role of sog protein is to bind to and antagonize dpp, the sog^A allele must be a loss-of-function mutation, whereas the sog^B allele must be a gain-of-function mutation. There are many ways to get a loss-of-function mutation, e.g. creation of a premature stop codon or a missense mutation that prevents any protein from being made. One way to get a gain-of-function mutation in *sog* would be for the mutation to occur in the region of the gene that encodes for the part of the sog protein that is recognized and cleaved by the protease tolloid. If the sogB protein is tolloid-resistant, then Sog protein levels will build up and restrict dpp activity.

4. Different mutations in conventional genes are noncomplementing, meaning that crossing two different mutant alleles does not rescue the phenotype. Although some mutational clusters in a complex locus can also be noncomplementing, they will also exhibit combinations of mutations that do complement each other. This is because complex loci have very elaborate arrays of cis-regulatory elements controlling gene expression in different tissues. Mutations that occur in a promoter or enhancer for one tissue can complement mutations that occur in the regulatory elements that control expression of the gene in a different tissue, as long as the two sets of regulatory elements function independently of one another.

5. *Ubx* normally controls the differentiation of segments T2P to A1. Loss-of-function mutations in *Ubx* result in the anteriorization of these segments (e.g. expansion of T1P/T2A segments at the expense of T2P-A1A). In other words, in the absence of Ubx, the more anteriorly-expressed homeotic genes of *ANT-C* are able to dictate the phenotype of the posterior thorax/anterior abdomen segments. In contrast, misexpression of *Ubx* anteriorly causes the anterior segments to adopt a posterior fate (parasegment 6), indicating the *Ubx* "dominates" over the *ANT-C* genes when co-expressed.

C. Extensions and applications

1. For each answer, the distribution of zygotic *hunchback* mRNA is shown on the right, and the distribution of caudal protein is shown on the left. Anterior is to the left, and posterior is to the right.

a. In a *Bcd^{null}* embryo there is no bicoid protein, so both gradients will be lost. The zygotic *hunchback* mRNA will be absent (since bicoid normally promotes *hunchback* expression), and the caudal protein will be found uniformly throughout the embryo (since bicoid normally represses caudal translation anteriorly).

 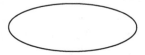

b. *Bcd^{77-489}* embryos have lost the YIPRL motif, and so the caudal protein will be found uniformly throughout the embryo. However, since the homeodomain is unaffected, the distribution of zygotic *hunchback* mRNA will be the same as wildtype.

c. *Bcd^{1-202}* embryos have the YIPRL motif, and so will show a normal gradient of caudal protein. However, because it lacks the homeodomain, the mutant bicoid will not be able to activate transcription of the zygotic *hunchback* mRNA.

d. *Bcd^{3X}* embryos express three times more bicoid protein than normal. As a result, there will be more *hunchback* mRNA and it will extend posterior much further, since bicoid promotes *hunchback* transcription. Conversely, there will be much less caudal protein, since bicoid represses caudal translation.

2.

a. As illustrated in figure 31.40, Hox genes are expressed progressively more posteriorly as you proceed from left to right along a Hox cluster. Thus, *Hoxa3* will most likely be expressed intermediate to *Hoxa2* and *Hoxa4* (for example, from approximately T1 to T6 or 7, although the exact pattern ultimately must be determined empirically).

b. Since loss of function of Hox genes usually results in anterior transformations, loss of *Hoxa2* will most likely cause a posterior expansion of C4-type vertebrae at the expense of C5 and C6.

c. Since overexpression of Hox genes generally causes posterior transformations, misexpression of *Hoxa4* in the region where *Hoxa2* is normally expressed would be expected to convert C5, C6 and maybe T1 to a more T5-type phenotype.

D. Standardized test format

1. A
2. D
3. C
4. A
5. D
6. C
7. B
8. C
9. C
10. D